This Far
by Faith

This Far by Faith

Stories from the
African American
Religious Experience

Juan Williams
and Quinton Dixie, Ph.D.

wm

WILLIAM MORROW
An Imprint of HarperCollins*Publishers*

THIS FAR BY FAITH. Copyright © 2003 by Blackside, Inc. All rights reserved. Printed in the United States of America. No part of this book may be used or reproduced in any manner whatsoever without written permission except in the case of brief quotations embodied in critical articles and reviews. For information address HarperCollins Publishers Inc., 10 East 53rd Street, New York, NY 10022.

HarperCollins books may be purchased for educational, business, or sales promotional use. For information please write: Special Markets Department, Harper-Collins Publishers Inc., 10 East 53rd Street, New York, NY 10022.

FIRST EDITION

Designed by Laura Lindgren Design, Inc.

Printed on acid-free paper

Library of Congress Cataloging-in-Publication Data
Williams, Juan.
    This far by faith: stories from the African American religious experience / Juan Williams and Quinton Dixie.—1st ed.
      p.   cm.
Includes index.
ISBN 0-06-018863-4
   1. African Americans—Religion. 2. African Americans—History.
I. Dixie, Quinton Hosford. II. Title.
BR563.N4 W513 2003
200'.89'96073—dc21   2002071884

03 04 05 06 07 ❖/RRD 10 9 8 7 6 5 4 3 2 1

In memory of Henry Hampton

CONTENTS

This Far
by Faith

INTRODUCTION

by Juan Williams

In Poland, beginning in 1981 as the union leader Lech Walesa battled to end Communist rule, Pope John Paul II gave his blessings and the power of the Roman Catholic Church to the freedom fighters. Polish himself, the pope offered advice, money, and even office supplies to strengthen the prodemocracy Solidarity movement. Then the pope traveled to Poland to defiantly praise the revolution that ended the Kremlin's hold on the country.

In South Africa Bishop Desmond Tutu, an Anglican, used his international pulpit to encourage sanctions on trade and investment until the government ended its racist policies of apartheid. In his purple clerical robes, and wearing a large cross around his neck, Tutu used the church as a shield as he defied South Africa's powerful police state.

In Central America Catholic priests have died for their brave defiance of regimes that ruled by military might. Oscar Arnulfo Romero, archbishop of El Salvador, became one of the few in the nation willing to denounce the military's use of violence and torture to silence political opposition. His piercing voice stirred opposition until he was assassinated in 1980, while at the altar conducting mass. But his willingness to defy the military and the sacrifice of his life, as he shared the bread of Christ, spurred a revolution and created new political leadership.

The pope, Bishop Tutu, and Archbishop Romero offer glittering examples of the power of religion to provoke and support social justice.

But God's power to transform society has no greater example than the U.S. Civil Rights movement. In America's battle with itself over slavery, legal segregation, and civil rights laws, the power of faith has been the cornerstone of efforts to save a nation's soul.

Unlike the pope or the bishops, who built their struggle with the help of an established church, the African American freedom struggle began outside any organized religion. As slaves, black Americans were stripped away from organized worship. They came to God not through the church but through faith.

Individual black people took on a cloak of faith, an unshakable belief that God would carry them through slavery and lift them up to freedom. Black people used the slave owners' religion to defeat slavery. They held up the Christian cross as evidence of their humanity. Black men and women used personal faith to claim a church that was founded by whites and initially antagonistic to blacks. With acts of faith they transformed themselves into Christian soldiers, marching for an end to segregation and for the promise of equality as God's children. Then they transformed the church. Black people built their own Christian churches. Those churches provided an organizational base that even the KKK and other racists feared to destroy because of the backlash prompted by any attack on a place of worship.

The influence of the church on the United States' Civil Rights movement stands out for the same reasons that made the church an agent of change in Poland, South Africa, and El Salvador. The church offered money, a protective structure for organizing, and leaders who used the skills they had honed in the pulpit to provide direction to the movement.

And in time the black church produced the leaders, the structure, and the vision for the greatest social movement in U.S. history, the Civil Rights movement that continues to this day. "Faith is the most powerful force in the world," in the words of African American theologian James

Cone. "It is the one thing. It is the light people can't put out." And at the base of that faith, according to Cone, is the belief among black Americans that "God has made us into a people not to be slaves but to be free." The scholar Cain Hope Felder identifies that same faith as a product of Christ-like suffering by black Americans. "Slavery and oppression," says Felder, "provide a wilderness experience where people raise ultimate questions and they are on the breaking point. They are on the brink. And at that special place between sanity and insanity, breaking or not is where I think God encounters us best. And I think at that point either you break or you get the sense that there's no need to break. And I think a lot of black people got the sense that there was no need to break." There was no need to break because they had faith.

Absolute faith and its power touched the life of Sojourner Truth and turned a slave woman into a champion in the fight to end slavery. It was faith as a point of community and culture that allowed Denmark Vesey to see himself as Moses come back to life to lead a slave revolt in South Carolina. It was faith that inspired young Henry McNeal Turner. As a child, Turner said, he had a mystical moment in which faith took over his body and dragged him down in the dirt so he felt the power of the Lord. The mystery of faith's power allowed him to see beyond negative images of black people and shout from the pulpit that God was a black man. The power of that idea was shocking at a time when the racist culture said any white man was superior to any black man and when Jesus, the son of God, was pictured only as a thin-lipped white man with long, straight hair. When white Christians refused to teach black children to read and write, much less allow them to attend the same schools as white children, black Americans used their own churches as a base to create schools for black children.

It was faith that led black men such as Elijah Muhammad, Noble Drew Ali, and Malcolm X to seek a non-Christian black religious experience. They felt a need to assert new, regal identities for black people despite daily insults and degradation. The African American faith experience is so big that it reformed Christian theology. All-consuming faith allowed for the creation of an inspiring political movement. Holding high the love Jesus showed for the poor and the weak, black leaders reached out to the masses of black and white people in the name of liberation and theology. Legendary intellectuals and activists such as Howard Thurman, James Lawson, Fred Shuttlesworth, and Wyatt Tee Walker made a new Christian church that preached the need for social reform.

The most notable of the nation's civil rights leaders was a child of the church, the Reverend Martin Luther King, Jr. When the nation's preeminent civil rights group, the NAACP, was outlawed in Alabama, King used the church's good name to form the Southern Christian Leadership Conference. King's fund-raising and public defiance of segregationist governors, mayors, and sheriffs were always clothed in the language of the scripture. Even on the night before he died King spoke in apocryphal terms of the Bible's "promised land."

In a crowded Memphis church King, a third-generation Baptist minister, gave thanks for living in the mid-twentieth century, when he could witness God's hand working from Johannesburg, South Africa, to Jackson, Mississippi, to free people. He said the "masses are rising up" and praised God for letting him add his hand to the sacred work of ending the bondage of black people in America. "I just want to do God's will," King said. Then, confronting fear of assassination, which took his life the next day, King reached back to the power of the Civil War, which brought about the end of slavery. He touched the chords of national memory by singing out: "And I'm happy tonight. I'm not worried about anything. I'm not fearing any man. Mine eyes have seen the glory of the coming of the Lord."

King's use of words from the "Battle Hymn of the Republic" took his audience back to a time when the organized Christian church was used by slave owners to perpetuate slavery. The only balm that the church offered to the broken souls of slaves was a poisoned vision of rewards in heaven for obedience to the slave master. In that time slaves found true faith to be the only fountain of hope for oppressed people. That faith was in God's spirit made flesh, Jesus Christ, as mighty liberator sure to free a people defenseless against the whip, exploitation, and abuse.

That deep well of faith from the darkest days of slavery sets the African American experience of religion apart. Christian faith sustained black men and women when their very humanity was being challenged. Faithful trust in God provided the basis for fighting slavery and defying Jim Crow laws of legal segregation and, finally, it provided a mighty weapon to disturb the conscience of white Christians. It is the faithful appeal to people of all colors for justice and equality—in God's name— that forever stands as the bulwark of the Civil Rights movement.

Religion in the African American tradition is still both a tool of survival and an inspiring "terrible swift sword" of justice. There is no separation between Sunday morning and the rest of the week. All life is spiritual and ever breath is full of faith in God as an all-powerful, ever-present force. It is God who sent the spirit to make the human body dance in joy; that sent the spirit to speak through true believers and affirm decisions in moments of crisis. It is God, and only God, who can confirm and redeem lost souls. How could any human slave master, no matter how long the reach of his stinging whip, compete with the master of the universe?

The depth of the black embrace of faith in even the most desperate of times, when millions died in the Middle Passage while being shipped to the New World, is reflected in the words of a hymn acknowledged as the Negro national anthem, "Lift Every Voice and Sing." Written by James Weldon Johnson, the first black man to head the NAACP, the song combines the spiritual legacy of deep faith with the fight for racial equality. The first verse reads, in part:

> *Sing a song full of the faith that the dark past has taught us,*
> *Sing a song full of the hope that the present has brought us.*

And in the third verse of this hymn, which was originally composed in 1900 to celebrate Lincoln's birthday, Johnson writes of long-standing trust between black people and the Almighty:

> *God of our weary years,*
> *God of our silent tears…*
> *Thou who hast by thy might*
> *Led us into the light,*
> *Keep us forever in the path, we pray.*

That sense of God's abiding involvement can be found in the music of the modern Civil Rights movement, too. In "We Shall Overcome," one verse says directly: "God is on our side, today." And in the refrain, people marching for equal rights and to protest the curse of racism drew strength from words that speak of faith in God's sense of justice: "Deep in my heart I do believe that we shall overcome someday."

Why did a people bruised by being sold in slavery and placed in a strange culture that degraded their humanity hold so fiercely to faith in God? By any ordinary measure, it is easy to view God as having abandoned these people. In fact, religion had been stripped, along with names and sometimes family, from slaves to render them isolated, weak, and unable to fight against slavery's oppressive harness. African tribal styles of worship as practiced by the Ibo, Yoruba, Fulani, and other ethnic groups were dismissed in this new world as primitive. White Christians looked down on African worship traditions, with its incantations, ancestor worship, and dancing. And when black people came to white Christian churches, there was little Christian grace. Blacks often had to sit in the back, in the balcony, or even outside. In some cases they had to attend separate services. This perversion of Christian fellowship only added to the difficulty of black people embracing Christianity. But somehow it did not stop them from believing that the God of Christianity was their God, too, and the problem was not with God, His church, or His word in the Bible. Faith said the problem was white racism.

Black people could have turned against the white Christian church, but instead they separated the message of Christian love from people who had no love for them. The impulse to join the American mainstream, to integrate and insist on equality as fellow children of God, proved greater than the anger sparked by racial rejection. This is the power of Christian love as interpreted by the struggle of African Americans. It is the power of belief in God to transform a world full of negativity, even hate, for people of color.

Even the Bible became a sharp sword to protect black people in their fight with racism. The story of Exodus, with Pharaoh holding the Jewish people in slavery in Egypt—and Moses leading his people to freedom by invoking the power of God even to part the sea—is a staple of the black pulpit. It inspired black people's faith in God's love for them and their struggle by giving them a model of God's help for another enslaved people. When slaveholders tried to use the Bible to instill servility in black people the slaves, many of whom had never been taught to read, told each other, "There must be another Bible within that Bible."

Dr. King, speaking in 1956, at the start of his ministry, addressed the segregationist distortion of the Bible. In a sermon based on an imaginary letter from the apostle Paul, King said black people understood the words of God in their heart and found that those who could read words in the Book were the ones suffering from an inability to understand. "I understand that there are Christians among you who try to justify segregation on the basis of the Bible," he said, speaking to the congregation as if he were Paul. "They argue that the Negro is inferior by nature because of Noah's curse upon the children of Ham. Oh, my friends, this is blasphemy. This is against everything that the Christian religion stands for. I must say to you as I have said to so many Christians before that in Christ 'there is neither Jew nor Gentile; there is neither bond nor free; there is neither male nor female, for we are all one in Christ Jesus.' "

King spoke for generations of black people who drew strength as well as comfort from the life of sacrifice and suffering led by Jesus. The example of Jesus Christ, God's own child, reaching out to redeem lost souls of all races and faiths was evidence that Christianity was the opposite of segregation. "Let them know," King preached, "that in standing against integration they are not only standing against the noble precepts of your democracy but also against the eternal edicts of God himself."

King and other civil rights leaders also engaged faith to transform black identity. The damage done by slavery and segregation twisted the self-image of many black people. They felt shamed by their darkness in a land where wealth and privilege attached to white skin. American culture equated black people with inferiority, stupidity, poverty, and every other evil. In faith's full embrace, however, black people stood as reflections of God—a divine creation. Negative associations with full lips, broad noses, and dark skin spun in reverse when faith entered the picture. Any negative judgment about black physical characteristics became evidence of white racists' limited vision. It was the racist who was flawed because he was blind to the variety of God's beautiful works.

W. D. Fard and Elijah Muhammad, following in the footsteps of Noble Drew Ali, used that logic to create new traditions, new identities, and a new sense of dignity for black people who followed them. Fard's unique theology included having him travel to make direct contact with God. In his orbit, black people were not only the equals of white people but people with special access to God and therefore white people's superiors.

That vision is too strong, too vindictive, for most. It is, however, the maximum dose of faith taken by people trapped in prisons, suffering in bad schools, and languishing in bad jobs. In a society with little confirmation for black men as vital, intelligent human beings, the power of faith in a just God to protect and comfort the oppressed is sometimes the only balm for terrible pain. It is faith tied directly to the power of God in the African American tradition of perseverance. In the darkest hour of personal doubt that kind of faith offers a reason to go on.

Truly amazing is the reach of faith across class lines in black America to this day. Faith leaps divisions of light- and dark-colored skin; it connects people living on country roads with people living on big-city avenues; and it connects black people in the church with black people in the mosque or temple. It is a binding force for the black American experience. Religious denominations and religious styles rise and fall. Black religious practices pour into the mainstream of American life with gospel singers becoming rock stars and rappers taking on the style of charismatic preachers. The unyielding center of the experience is faith, faith that God will guide and protect.

The gospel song "We've Come This Far by Faith" includes these lines:

We've come this far by faith. Leaning on the Lord.
Trusting in His holy word. . . .
Just the other day I heard someone say he didn't believe in God's word.
But I can truly say that God has made a way. And He's never failed me yet.
That's why we've come this far by faith.

In the pages that follow, faith creates new life.

"God Has a Hand in It"

The God of Bethel heard her cries,
He let his power be seen;
He stopped the proud oppressor's frown,
And proved himself a king.

—RICHARD ALLEN

Charleston, South Carolina, today is a coastal city best known for its colonial past. The city is filled with exquisite homes built in the 1700s for wealthy plantation owners, shipbuilders, and importers. Tourists come to Charleston to see these grand houses, with their large, leafy gardens, elegant fountains, and porches with grand white pillars, which hint at the fabulous society life that once thrived here. All around the city are churches established before America was a nation. And at the center of the city is the Slave Market, which is now part museum and part crafts fair. But that slave market was the economic lifeblood of Charleston in the 1700s. It is the place where three-quarters of all slaves entering the United States first set their feet in America.

In the late 1700s so many slaves were arriving that Charleston was a densely populated, mostly black city. Aristocratic whites were a

Slave auctions were common in Charleston
and were as much social events as business transactions.

Charleston was a thriving port city throughout the nineteenth
century and had a small yet significant number of free blacks
at the time of Denmark Vesey's planned rebellion.

small population who ruled over the black slaves, both on the enor-
mous plantations toward the coast, which were the foundation of
Charleston's economy, and in the heart of the city. Rice and cotton, the
region's key cash crops, were labor-intensive and thus required a large
number of field hands. So low country planters had been ignoring the
1808 United States ban on the international slave trade for years, and
they continued to import Africans, whom they felt were familiar with
rice cultivation or better suited for the backbreaking labor of cotton
picking. From 1810 until 1900, South Carolina had a black majority,
with its black population reaching nearly 60 percent on the eve of the
Civil War. And in 1822, in the heart of this South Carolinian city, in a
small home not far from the Slave Market that still exists, a group of
thirty men huddled near the hearth, as one drew diagrams on the dirt
floor, plotting a rebellion in urgent, hushed tones.

The room was hot and stuffy, and it had the thick smell of too
many people who had been crowded in the humid summer heat for

SLAVE TRADE DATA

Any effort to identify the specific homelands of the African slaves brought to the Americas from the fifteenth through the nineteenth centuries will be a daunting task. Technological advances have helped widen the base from which historians and students can retrieve information, but the actual number of slaves exported from Africa as well as their origins remains ambiguous and difficult to interpret. Nearly all data on the slave trade are estimated figures.

The estimated number of Africans exported from Africa ranges from 8 million to 25 million. Recent studies have projected that the transatlantic slave trade transported anywhere from 9 million to 15 million Africans. These figures do not include the Africans who perished during the journey across the Atlantic. The first Africans entering the slave trade left the continent in 1441 on a Portuguese boat, and the first Africans to land in North America arrived in Virginia in 1619.

**ESTIMATED NUMBERS OF AFRICANS
IMPORTED TO THE AMERICAS, 1451–1870**

From Christina K. Schaefer, *Genealogical Encyclopedia of the Colonial Americas*
(Baltimore, Md.: Genealogical Publishing Company, 1998)

British North America	532,000
Spanish America	1,687,100
British Caribbean	2,443,000
French Caribbean	1,655,000
Dutch Caribbean	500,000
Brazil	4,190,000
	11,007,100

AMERICAN DESTINATIONS OF SLAVES BY REGION IN AFRICA

From Genealogical Encyclopedia of the Colonial Americas

REGION IN AFRICA	DESTINATION IN THE AMERICAS
Angola	Barbados, Brazil, Cuba, Haiti, Jamaica, Mississippi, South Carolina, Virginia
Gold Coast (Ghana)	Jamaica, then on to South Carolina, Virginia
Mozambique	Cuba, Haiti, Jamaica, Mexico, Peru,
Nigeria	Brazil, British Guiana, Cuba, Haiti, Jamaica, St. Croix, St. Thomas
Northern Nigeria	West Indies
Senegambia (Senegal and Gambia)	Barbados, Brazil, Cuba, Mississippi, South Carolina, Virginia
Sierra Leone	Brazil, Cuba, French Guiana, Guadeloupe, Haiti, Jamaica, Martinique, Mexico, Mississippi, Peru, Puerto Rico

NUMBERS AND PERCENTAGES OF SLAVES
EXPORTED FROM AFRICA TO THE NEW WORLD

Compiled from David Northrup, *The Atlantic Slave Trade*
(Boston: Houghton Mifflin, 2001)

	NUMBER	PERCENTAGE
1450–1600	367,000	3.1
1601–1700	1,868,000	16.0
1701–1800	6,133,000	52.4
1801–1900	3,333,000	28.5

SLAVES IMPORTED TO NORTH AMERICA

From Philip Curtin, *The Atlantic Slave Trade: A Census*
(Madison: University of Wisconsin Press, 1969)

COASTAL REGION OF AFRICA	To Virginia 1710–1769 %	To South Carolina 1733–1807 %	British Slave Trade 1690–1807 %	All Imported to N.A. %
Angola	15.7	39.6	18.2	24.5
Bight of Benin	——	1.6	11.3	4.3
Bight of Biafra	37.7	2.1	30.1	23.3
Mozambique-Madagascar	4.1	0.7	——	1.6
Senegambia	14.9	19.5	5.5	13.3
Sierra Leone	5.3	6.8	4.3	5.5
Windward Coast	6.3	16.3	11.6	11.4
Unknown	——	——	0.6	0.2

too long, bringing the sweat of their day's labor with them. The fear was palpable. Eyes darted around the room, someone stood by the door, and a lookout outside kept a close watch for inquisitive whites. One of the men, the governor's most trusted servant, had already offered to kill the governor and his family as they slept. Someone else was coordinating the distribution of weapons in the days before the uprising. The two men closest to the light, their faces dancing with shadows, were the leaders of the rebellion. The man next to the fire was Denmark Vesey, a class leader at the nearby church and the most outspoken black man in the city. And standing just behind him was the most powerful religious leader in Charleston's rural slave plantations, Gullah Jack Pritchard.

Vesey was over fifty-five: a life span longer than most blacks in South Carolina could expect. He was a tall, dark-skinned man with graying hair and missing teeth. His hands were leathery from a life of

TO BE SOLD on board the Ship *Bance-Island*, on tuesday the 6th of *May* next, at *Ashley-Ferry*; a choice cargo of about 250 fine healthy NEGROES, just arrived from the Windward & Rice Coast. —The utmost care has already been taken, and shall be continued, to keep them free from the least danger of being infected with the SMALL-POX, no boat having been on board, and all other communication with people from *Charles-Town* prevented. *Austin, Laurens, & Appleby.*

N. B. Full one Half of the above Negroes have had the SMALL-POX in their own Country.

Slave advertisements gave prospective buyers a general sense of the cargo's place of origin so they might determine slaves' familiarity with certain kinds of agricultural production. Over time stereotypical behavioral traits came to be associated with African slaves depending on their regional and cultural background.

work as a carpenter. His strong physical presence, as well as his stature as a leader in Charleston's black church community, led many white people to label him a rabble-rouser who was best avoided. In fact, Vesey was known for refusing to yield the right-of-way to any white man as he walked on Charleston's narrow sidewalks. He even refrained from bowing to the white aristocracy. These small acts of defiance were just outward signs of Vesey's conviction that, regardless of his skin color, he was not to be treated as anything less than equal to any man in God's sight.

Denmark Vesey had had to live by his wits from a young age; his faith in God and his mind were all he had to steady him in the face of oppression and abuse. At age fourteen, he was sold to a sugar plantation in Haiti where enslaved Africans were expected to chop cane from sunup to sundown. To avoid work, the boy spent the bulk of his time sprawled on the ground, claiming to be a victim of epileptic seizures. White doctors in the town confirmed the boy's illness, and when the captain of Vesey's slave ship returned to Haiti three months later, he refunded the plantation owner's money and took custody of this sick slave boy. The captain assigned Vesey to be his personal assistant. Perhaps it was Divine Providence, or perhaps it was the sea air that miraculously cured Vesey of his ailments, but Denmark Vesey never again displayed signs of epilepsy.

Religion was a form of resistance among slaves, and woven throughout the narrative of African peoples' enslavement in North America are practices, beliefs, and traditions deeply rooted in West African religious philosophies. Unlike Europeans, for whom religion is neatly tucked into a particular category in society, West Africans put religion and God in particular at the center of life. It was normal behavior for slaves to cry out to the West African gods for assistance in their struggle for liberation. Their cries took various forms: drumming and dancing, veneration of the ancestors, spirit possession, ring shouts, African-style burials, dancing, and ritual sacrifices. The historian Sterling Stuckey argues that memory allowed slaves to retain these practices and beliefs, which in turn empowered them to resist slavery and enabled the African gods to remain visible and active in slave culture. African religious philosophy clearly manifested itself in the Americas through three major systems of belief: Voodoo, Santeria, and Candomblé.

Yoruba religions, from which emerged traditions such as Voodoo and Santeria, are arguably the most critical and recognizable African-based traditions in the West. Yoruba contains an elaborate cosmological system with more than four hundred deities known as *orishás* and an overarching deity named Olorun (owner of the sky). In addition, Yoruba philosophy places a significant importance on remembering the family's ancestors. Ogun is one of the most important deities. The god of iron making and of war, Ogun is the deity for blacksmiths, warriors, and those who use metal in their occupations.

Voodoo is one of the most popular African-influenced religions in the Americas and arguably the least understood. Originating in Haiti, Voodoo illustrated, according to the historian Robert Farris Thompson, a "signal achievement of people of African descent in the western hemisphere: a vibrant, sophisticated synthesis of the traditional religions of Dahomey, Yorubaland, and Congo with an infusion of Roman Catholicism." In the United States Voodoo flourished in New Orleans and was primarily associated with slaves from Haiti. In Voodoo, practitioners follow different spiritual paths and worship a pantheon of spirits called Loa, meaning "mystery." The Loa are similar to Christian saints; indeed, they are individuals who led exceptional lives on earth. Followers of Voodoo believe that each person has a soul that contains two parts: a *gros bon ange,* or "big guardian angel," and a *ti bon ange,* or "little guardian angel."

Santeria also originated from the West African religions. It has one primary God, known as Olodumare or Olorun, which represents the source of *ashé,* the immanent and transcendent spiritual energy of the universe and all of its creatures. Olodumare manifests itself and communicates to the world via the *orishás,* which can rule over nature and all aspects of human life. In fact, the followers of Santeria often approach the *orishás* for assistance in times of distress and difficulty. Enslaved Africans communicated with the *orishás* through prayer, ritual, divination, and sacrificial offerings, and in most cases these rituals fostered the fervor needed to become possessed by one of the *orishás*. During possession one often revealed prophetic messages from the *orishás* to the people. In the Americas these practices, especially the appropriation of the *orishás*, were veiled behind Roman Catholicism.

Candomblé, which emerged in Brazil, revolves around rituals of sacrifice, singing, spirit possession, and drumming. Practitioners praise and sing songs to the *orishás* to gain favor and recognition from the gods.

Africa's religious and philosophical traditions survived the Middle Passage, and when the Africans landed in the New World, their gods began manifesting in numerous ways. Sometimes African religious themes were veiled, other times they were fully exposed within slave culture. Regardless of how Africans expressed their religious beliefs and practices; the prints of African philosophy were and remain visibly clear in black religious traditions.

The African Church was built after blacks withdrew from the biggest Methodist church in the city. In the early nineteenth century, African religions were an important thread in South Carolina's religious tapestry, particularly on the coast, the so-called low country. With new slaves arriving regularly, there was a constant source of renewal for these traditions. The relative isolation of rural plantations and the large slave populations enabled enslaved Africans to keep their own beliefs alive, and the potent spiritual powers that so many of them believed in stood as testaments to the resilience of African religious traditions. With so many religions available to them, blacks tended to borrow bits and pieces from various traditions, creating a religion unique to them. Even Christianity was a part of their fusion of faiths, and it was common for Charleston-area blacks in the 1820s to belong to one or more local faith communities.

It was in 1794 that the first independent black Methodist church came into being. Ushers at the predominantly white St. George's Methodist Church in Philadelphia pulled black worshipers from their knees during prayer, and the incident provided the necessary momentum for the city's black Christians to found their own churches. In 1794 Philadelphia's Christian black community decided to form an Episcopal congregation, which they called St. Thomas African Episcopal

As pastor of St. Thomas African Episcopal Church in Philadelphia, Absalom Jones was the first Episcopal priest of African descent in the United States.

Church. Absalom Jones, leader of the group, became the first black Episcopal pastor in the United States.

Richard Allen and the rest of the black Philadelphia Methodists, however, were in a bind. Righteous indignation would not permit them to return to St. George's Church, but the city's presiding elder refused to recognize them if they formed their own church. Nonetheless, Allen informed the white Methodist officials, "If you deny us your name, you cannot seal up the scriptures from us, and deny us a name in heaven. We believe heaven is open to all who worship in spirit and truth." With that declaration of independence, he led the remainder of St.

George's dissidents to form the Bethel Church. By 1816 Allen and his congregation had attracted the attention of blacks and whites all along the eastern seaboard, as well as in the slaveholding South. Blacks in Baltimore and other cities experienced segregated Sabbaths, as well as threats to their quest for autonomy. So late in 1816 Allen called together delegates from black Methodist communities throughout the region and they formed the African Methodist Episcopal Church.

The audacity of the Northern AMEs disturbed the white elite. Not only did they challenge white authority in open court but they went so far as to claim God had

Richard Allen was the first pastor of Bethel AME Church in Philadelphia and the first bishop of the African Methodist Episcopal denomination. He is often given credit for igniting the black independent church movement in America.

granted them victory. What would happen if Southern black residents followed suit? Of course, Charleston's African Church probably would not have annoyed local whites that much if it had engaged solely in hymn singing and testifying. For whites in the city were accustomed to enslaved blacks maintaining their own cultural practices, and they permitted such pursuits as long as they did not attempt to unravel the city's delicate social fabric. But the church's members and leadership were committed to a model of Christian community that did not segregate sacred and secular concerns. Their vision of God's will led them to care for one another's material, as well as spiritual, welfare. So they pooled their resources and built a church structure, bought a burial ground, occasionally even purchased enslaved congregants' freedom. Whites felt that such boldness was only enhanced by affiliation with Philadelphia's black Methodists, and they were determined to put an end to it.

On December 3, 1817, the Charleston city guard invaded the church and arrested 469 blacks on charges of disorderly conduct. Whites in the vicinity had apparently complained to authorities about the Africanized "species of worship" conducted by the ministers. Undeterred, black worshipers petitioned the South Carolina House of Representatives in 1818 for permission to maintain a separate house of praise. In part, the petition read:

> The free persons of colour attached to the African Methodist Episcopal Church in Charleston called Zion, have erected a house of worship at Hampstead on Charleston Neck at the corner of Hanover and Reid Streets. Petitioners request to open said building for the purpose of Divine worship from the rising of the sun until the going down of the sun.

In an attempt to assuage the white assembly's fears, the elders of the church assured them that "white ministers of the gospel of every denomination shall be repeatedly invited to officiate." Instead of approval, the petition gained only the white community's wrath. Some couldn't understand why blacks insisted on holding separate services, since "it is well known," a Charleston newspaper pointed out, "that in every church in Charleston and throughout the state, accommodations are provided for such negroes and free people of colour as choose to attend." The problem, however, was that the accommodations in those sanctuaries were segregated and forced blacks to worship as second-class citizens.

Despite, or perhaps because of, the showing of support for the church by Northern black churches, the rage and fear of local whites only grew. On the first Sunday in June 1818, the city guard raided the African Church again. This time they singled out the congregation's leadership. They arrested nearly 150 people, mostly ministers, elders, and class leaders, for violation of city ordinances. Some were given the option to pay a five-dollar fine or submit to ten lashes at the workhouse, while those whom whites viewed as key to the birth of the black church received sentences of either one month in prison or banishment from the state. This new tactic was clearly aimed at decapitating the congregation in the hope that it would wither away. But, unwilling to abandon their dreams, the church's leaders chose to serve the prison sentence in order to stay with their flock.

There is no record of exactly who was arrested, but it is almost certain that Vesey, by then one of the best-known black leaders in Charleston, was among them. This experience was a turning point in his philosophy. From then on he firmly believed that rebellion was the only way blacks could ever truly gain freedom. He was not alone in his conviction. During his confession in 1822, Smart Anderson reported that one group of slaves wanted to rise up when the whites arrested the ministers and class leaders. But the time was not yet right. There was a great deal of preparation required if blacks were going to cast off the yoke of oppression. Vesey found a few men he felt he could trust to follow through with him to the end. Yet before they could reach the moment of the meeting in 1822, they would need munitions, money, manpower, and, most important, Divine Providence.

From the beginning Vesey believed that God was on the side of the enslaved blacks and that scripture was a supporting witness to his claims. He started out as a member of the white Second Presbyterian Church, but he soon switched to the African Church, prompted by a

ORIGINS OF THE
BLACK BAPTIST MOVEMENT

The emergence of independent black Baptist churches in the United States in the eighteenth century occurred during the effort by American colonialists to eradicate British rule in the colony. On the periphery sat two critical cultural phenomena: the rise of "revivalistic" congregational churches and the ongoing efforts to eliminate slavery. From these efforts came the critical means for blacks not only to fight for their liberation but also to build their own institutions. Indeed, the War for American Independence created the philosophical framework from which blacks argued that liberation was intended for *all* people. The Revolutionary War enabled blacks to appropriate the "master's tools" for the dismantling of racial oppression. In addition, the rapid expansion of congregational churches opened doors through which blacks could take control of their religious life and build visible institutions that would foster their struggle for emancipation.

Two significant architects of the independent black church were George Liele and David George, both former slaves and self-educated ministers. Liele, born into slavery in Virginia in 1750, was one of the first blacks to be ordained a Baptist preacher in the United States. Unlike many slaveholders, Liele's owner, Henry Sharp, supported his interest in the ministry. Ordained by the Buckhead Creek Baptist Church in Burke County, Georgia, Liele was commissioned to preach the gospel to slaves on nearby plantations. He eventually gained his freedom and in 1777 helped establish the first African Baptist Church outside Savannah, one of the first black Baptist churches in the United States. After the British evacuated Savannah in 1782, Liele headed for Jamaica. He indentured himself to the British in exchange for a boat ride to the island, where he eventually paid off his debt and once again began preaching. But in Jamaica, a stronghold of the Anglican Church, Liele faced opposition from Anglican officials. His efforts to build an independent black church soon landed him in jail. The ostensible crime? He had acquired too much debt while building a church in Kingston. After a short period in jail, Liele returned to preaching. His church's membership ballooned, and he baptized some four hundred people before his death in 1820.

David George, who had been owned by Indians, was mentored by Liele for the ministry. The reluctant George eventually helped form a Baptist church on a plantation in Silver Bluff, South Carolina. Under his leadership the church grew to thirty members, but its growth was soon stunted when the British offered freedom to blacks in exchange for their work and support of the British in the war. George took advantage of the offer and supported the British until 1782, when he relocated to Nova Scotia in the hope of acquiring full freedom. But when he arrived he faced all types of discrimination, in particular opposition from whites in Shelburne, where he attempted to build a church. As people became familiar with his ministry, the resistance to him waned, but it never quite died. So when the governor of Nova Scotia, John Clarkson, presented George with an opportunity to evangelize in Sierra Leone, he enthusiastically accepted. In 1792 George established a Baptist church in Free-town, Sierra Leone. His ministry stretched into all aspects of social and political life in the country; indeed, he led efforts against British rule, such as the exorbitant taxes on the people of Freetown. When he died in 1810, George's church had two hundred faithful members.

Although Liele and George played critical roles in the origins of a vital institution that would give blacks in the United States important social power, the historical context prevented both of them from building their ministries here.

desire to maintain his dignity while at worship—something that was impossible in segregated pews. Vesey was a class leader at church, and he was therefore responsible for the spiritual well-being of a select group in the congregation. It was during class meetings that he emphasized God's commitment to those on the bottom rung of the social ladder. Indeed, class meetings were both his theological proving ground and his base of recruitment. He relied on claims of scriptural mandate to convince his audience to join the rebellion.

Sometimes he would address groups of men on local plantations, and other times he would try to persuade those he encountered on the streets of Charleston. Either way, his message was the same: Almighty God wants his people to be free, and if God be for us who can be against us? Although many black religious leaders drew from the New Testament to teach their congregations, Vesey preferred the Old Testament image of God the conqueror, who would bring his people forth from slavery with an outstretched arm. He did not use religion to urge people to stay in their place, turn the other cheek, and await salvation in the life to come. Rather he encouraged members of his class to take a stand for just treatment in this life. He often drew on the Book of Exodus and its chronicle of the Hebrews' bondage, suffering, and ultimate liberation.

The parallels between the saga of the Hebrews and that of enslaved and denizen blacks served Vesey's needs perfectly. Rolla Bennett, one of Vesey's lieutenants, later recalled that during the secret late-night meeting at his home, Vesey read to the men assembled about the deliverance of the children of Israel from Egyptian bondage. William Paul confirmed that Vesey "studies the Bible a great deal and tries to prove from it that Slavery and bondage is against the Bible." One of his favorite proof texts was Exodus, Chapter 21, which, among other things, outlines God's commands on the treatment of slaves. Verse 2 states that a Hebrew could enslave another Hebrew for only six years; in the seventh year he and his family must be set free. Often Vesey read to potential recruits Exodus 21:16, which states, "He that stealeth a man, and selleth him, or if he be found in his hand, he shall surely be put to death."

Vesey's followers hung on his every word, affirming that God would be with them just as he had been with his people in the days of old. One slave, Abraham, wrote to a fellow slave, Peter, of his desire to join the liberators. "With pleasure," he said, "I will give you an answer. I will endeavor to do it. Hoping that God will be in the midst to help his own. Be particular and make a sure remark. Fear not, the Lord God that delivered Daniel is able to deliver us."

Vesey used much the same approach in recruiting Jack Pritchard, who was both a religious and a political leader to slaves working on plantations outside Charleston. These rural slaves were quite distinct from the craftsmen, servants, clerks, and laborers that populated the city. A nervous Pritchard kept touching a talisman around his neck as he made his way to that late-night meeting at the rectory. Dressed in the dark clothes of a country dweller, in woolen pants and a black vest, he nodded politely to whites as he walked. He even stopped at one point when a white man approached him and stood to one side until the man had passed by. Unlike Vesey, Pritchard had no problem with deference to whites.

Despite his outward affectation of humility, folks who knew Jack Pritchard held him in awe. He was a leader—a man who used the fears, desires, and elements of African religion, including Voodoo, to elevate himself among the blacks working on the plantations around Charleston. Folks claimed he was a conjurer—a roots doctor who possessed magical power that could either help or harm those at whom it was directed. He saw himself as a guardian of the ancient practices of the old country, and he trained others to become practitioners so the religious fabric of the ancestors might live on. Pritchard knew that it was his spiritual influence over the Gullah community that made him valuable to Vesey. Without the power of God at their disposal, the liberators' thoughts of freedom would quickly fade like a fistful of Jack's magic dust.

Short, stout, and balding, with a long, thick mustache, Gullah Jack, as he was called, did not seem physically threatening. But he had a unique influence over the blacks of the low country—the Gullahs. These people spoke a language, and practiced a culture, that was almost wholly inaccessible to whites. Among them were not only African blacks but some blacks from Haiti, who wound up in South Carolina when their masters fled the island in the wake of Toussaint-Louverture's revolution. They added French and Creole to the Gullah mixture, not to mention firsthand experience of what was possible if blacks rose up as one unit in defiance of white tyranny. Moreover, these Haitians carried to the United States their Voodoo beliefs, which thickened the blend of enslaved African religion in South Carolina. Although it does not seem that Gullah Jack knew French or Creole, he undoubtedly recognized the core of Voodoo rituals and, quite possibly, extended his influence among this population by incorporating some of their practices into his belief system.

Outsiders could understand only what the Gullahs wanted them to understand; the rest remained hidden beneath a cultural blanket that no white person could penetrate. This secrecy proved to be a key weapon of resistance for low-country enslaved blacks. But Gullah Jack was different even from most of the Gullahs, for he could move seamlessly between the rural and urban communities. Vesey knew that he needed Gullah Jack for his rebellion to succeed, and he wasted no time in making Jack Pritchard his head lieutenant.

No one is sure where the word *Gullah* comes from. In the early colonial period plantation owners in this region showed a preference for slaves from the Ngola people (from near present-day Angola), and some say that the word *Gullah* is shorthand for this group's name. But throughout the nineteenth century low-country slaves overwhelmingly came from the Windward Coast region of Africa (present-day Liberia and Sierra Leone). The Gollas lived in that area, and many scholars think this ethnic group provided the base for Gullah culture. The truth probably lies somewhere in the middle, for nineteenth-century Gullah life, language, and religion represented a mixture of European and many native African practices as enslaved Africans adapted to life in a strange land. Regardless, Gullah Jack knew who he was—a man guided and protected by the spiritual powers of his God—a spirit far superior to all the evil present in the world.

Pritchard knew that the rebellion would need manpower if it was going to be successful, but he was shocked to see the more than thirty men crowded on the floor of Vesey's home when he arrived at the meeting. He was not sure it was wise to invite so many to a strategy

South Carolina holds a significant place in the history of African peoples in the United States in general and the development of the African American religious ethos in particular. In this Southern colony a community of African Americans maintained and fostered cultural practices and belief systems from their African heritages, such as the identification of religion as central to all aspects of their culture and social existence. Blacks began to constitute the majority of South Carolina's population in 1708. And their isolation from Europeans provided a milieu in which African cultural practices and beliefs could be retained.

Many of the people transported to South Carolina from Africa could trace their roots to the trading posts in the Congo-Angola region and the land between the Senegal and Gambia Rivers in West Africa. For a significant period of antebellum history, South Carolinian blacks were born in Africa and upon their arrival in America were confined to the coastland. This was where historians locate the Gullah people, who clung to their African culture and made it possible for historians and students of black history to understand how slaves integrated their African pasts with their New World experiences. Indeed, the Gullah people maintained their own dialect and unique religious practices. One historian argues, "Slaves converted Christianity to their culture."

From the beginning of South Carolina's history, the Anglican Church had a strong hold, but the Protestant tradition did not attract many black slaves. For one thing, participation in the Anglican tradition required literacy—members had to memorize the Lord's Prayer, the Apostle's Creed, and the catechism—but the government prohibited education among the slaves. However, the South Carolina Methodist federation became the first religious tradition openly to advocate religious instruction among slaves, which, of course, required some form of literacy instruction. Hence, when the Methodist tradition rose in popularity throughout South Carolina, thanks to the eighteenth century's Great Awakening and the evangelism of George Whitefield, it grew especially in popularity and prominence among African Americans.

Despite the influence of Christianity on the customs and beliefs of slaves, West African rituals and philosophies remained deeply interwoven in their religious practices. The overarching theme of African religious belief is the recognition and celebration of community. African religious practices were retained in three areas of slave life. First, evangelical Christianity provided the framework for slaves to maintain the West African tradition of secret societies. Like secret societies, the church had an informal (and often unspoken) initiation process. Before officially joining the church, slaves had to testify about their conversion. If the elders accepted their testimony, the slaves were baptized and thus received full membership in the church. Status as a member of the new community was marked by water ritual both in Africa and in South Carolina.

Second, African retentions were evident in how slaves conceptualized and responded to death. For Africans, death symbolized a rite of passage, the mark of entrance into the next world. The dead person's spirit remained alive and vital in his or her community. For instance, slaves decorated burial sites with personal items and lamps and torches, which would guide the spirits to the next world.

The ring shout constitutes another African religious retention. The Gullah people created Praise Houses, unassuming, paintless buildings that resembled small barns. There slaves fostered community and fully expressed themselves in, among other activities, the ring shout. The ring shout involved altered states of consciousness that resembled spirit possession or what Christians called being filled with the Holy Spirit. Groups of people would stand in a circle, clap their hands, and swing their bodies from side to side, while others would scrape their feet on the floor.

The rich presence of African elements in slave culture in South Carolina coincided with an unusually high number of attempts to disrupt slavery, including the Denmark Vesey Rebellion in 1822 and the Stono Rebellion of 1739. In fact, state officials contemplated ending the Christianization of slaves because many insurrections emerged from independent black churches.

session. What he did not know—what he couldn't have known—was that their downfall would come from a source altogether outside this meeting.

In addition to Pritchard and Vesey, the clear leaders of the meeting, the assembled leaders of the plot included Ned Bennett, Rolla Bennett,

Monday Gell, and Peter Poyas. Ned Bennett was a slave for Thomas Bennett, South Carolina's governor, who lived just down the block from Vesey. He volunteered to lead the non-French-speaking slaves in the country outside Charleston. Monday Gell ran a shop for his master making harnesses. His contribution was access to guns and an ability to cultivate support from Charleston slaves who came from Africa's Ibo tribe, as he did.

Vesey's cluster of confidants was just as revealing for whom it excluded as for whom it included. Conspicuously absent was Morris Brown, leader of Charleston's black Methodist community. Vesey thought that as a free man of mixed parentage, Father Morris, as he was called, had too much to lose to be included in the inner circle. He never gave any indication of why Brown was not consulted, although Gell later told friends that Vesey thought Brown might turn them in. By the same token, it is quite plausible that Vesey kept Brown in the dark because he wanted to protect him and the African Church. After all, the city fathers had made it clear that they were looking for a reason to drive Brown out of town.

Also missing from the rebellion's leadership and recruitment lists were members of a group called the Brown Fellowship Society. Unrelated to Morris Brown, the Brown Fellowship Society was a social organization established in 1790 and composed of Charleston's mulatto elite, who wanted both blacks and whites to know that they were different from common Negroes. Undoubtedly some of them shared lineage with local whites, and Vesey believed he could not rely on them to support his efforts. In fact, he tended to avoid association with free blacks, many of whom felt they had little in common with the enslaved population. When William Penceel heard about the planned rebellion, he declared he would "have no part in such business" because he was already free.

Vesey and his lieutenants maintained a pattern of evasion when it came to browns. Like Vesey, some browns were members of the African Church, but overwhelmingly they maintained membership in the predominantly white congregations in Charleston. The Brown Society did not see a need for a separate African house of worship, nor did they approve of the Africanized Christianity expressed in that congregation's music, dance, and preaching style. In contrast, Vesey and others preferred the African Church to the incessant humiliation in a white-dominated House of God. Furthermore, complete separation allowed blacks the opportunity to pray, teach, learn, and praise the Lord in a way that suited them.

As the group trickled out of the house, they could feel excitement and fear. Their eyes shone with thoughts of the fires they would light on the night of the rebellion. They imagined the chaos that would ensue, leading to the slaughter of every white man, woman, and child who crossed their path. Blacks so drastically outnumbered whites in Charleston that the group felt a tense optimism. After years of slavery and oppression, they were finally, cautiously, hopeful that the tide could be about to turn in their favor. Perhaps it was this sense of destiny that led Denmark Vesey to approach a well-known slave named Joe on the street one afternoon.

The more successful Vesey's recruitment efforts were, the more brazen he became in approaching rebels. So when Joe meandered by on his horse that afternoon, Vesey didn't think twice about grabbing the horse's reins and holding it in place until he was done with Joe. He asked Joe how he felt about being a slave, and Joe, ever eager to turn the other cheek, replied, "Well, if God wills me to be a slave, then I don't suppose there's much to be done about it."

Vesey told Joe when Hercules found his wagon stuck in a ditch, he prayed to God, asking the Almighty to do something for him. According to Vesey, God said, "You fool, put your shoulder to the wheel, whip up the horses, and your waggon will be pulled out!" For Vesey, God was always a help in times of danger and struggle, but he helped only those who were willing to help themselves. Vesey went on to describe the uprising, now just weeks away, to the silent Joe. And when he finished, confident that he had won another fighter for the cause, he released Joe's horse's reins and sauntered away.

Joe didn't know what to do. He had always been taught that God abhorred violence, regardless of whether humans used his name to justify it. To Joe everything that Vesey said went against the lessons of the New Testament. Joe knew that you were supposed to turn the other cheek, treat your neighbor as yourself, and rejoice when persecuted for the sake of righteousness. The God he worshiped could never condone the slaughter of whites, even if they had oppressed blacks for years. Joe didn't want to turn Vesey in, but his conscience wouldn't let him stand by and watch innocent people die. So he went to the most pious person he could think of: George Wilson.

Wilson was lauded by both whites and blacks as a man of character. Like Vesey, he was a class leader in the African Church, and he worked tirelessly to spread the message of love and mercy that he believed God wanted him to convey. Joe saw no way to reconcile Wilson's God of love with Vesey's God of wrath. Racked by guilt and indecision, Joe told Wilson everything he knew—that Vesey, the charismatic carpenter, was plotting a rebellion that promised to end the lives of tens, if not hundreds, of whites in the city. He told him that Vesey was claiming to have God on his side and that he didn't know which way to turn. Joe was afraid for the lives of blacks if he told, and afraid for the lives of whites if he didn't. He was afraid of betraying Wilson's God, and he was afraid of the wrath of Vesey's God. Wilson listened carefully, and when Joe was done he told him that he'd done the right thing.

Our loving God would never condone violence, he said, and the rebellion had no chance of success. The only way to prevent violence would be to inform the whites of the plot.

Major John Wilson was the first white to know. As George's owner, he trusted his slave completely and knew him to be telling the truth. Major Wilson knew that his life and the lives of his friends and family were in danger, but he needed more information before he could take real action. So at the major's request George Wilson became a spy for the white authorities. He used his reputation as a man of character to find out everything he could, and with every fact that his friends passed on to him, the fate of the rebels became more firmly sealed.

George approached his friend Rolla Bennett and asked him about the uprising, saying that he wanted to protect himself from the possible violence. Bennett, sympathizing with his long-trusted friend, told him in a whisper that if he wanted to be safe he should leave town on the second Sunday in June. Wilson was overcome with guilt at the knowledge that he was going to betray his friend. On the verge of tears, he pleaded with Bennett to stop the uprising before anyone got hurt. Bennett just cracked a smile and said, " 'Tis gone too far to be stopped now."

Late at night on June 16, 1822, the air was thick with adrenaline. Underneath the sounds of cicadas and grasshoppers, people rustling in the underbrush crept toward Charleston. Vesey's recruits were readying themselves in their homes even as Gullah Jack's followers converged on the outskirts of the city. Each one had eaten groundnuts and parched corn that morning, and each held a *cullah*, or crab claw, in his mouth now. They believed that with Gullah Jack's charms they couldn't be hurt. The slave owners' bullets would whiz past them, and they would enact the carnage and vengeance that they had been imagining for so long.

One of them, crouched in the bushes, spotted what looked from a distance like a phalanx of men. But it couldn't be. The whites were all asleep, and God was on their side. On the other side of town, another man spotted the same thing. And another. As they crept closer, the rebels realized that not only was the city guard out in full force but they were armed, prepared for every one of the rebels.

The rebels paused while the full impact of what had happened sunk in. Their element of surprise was gone. Their assured victory had disappeared. They could blaze forward into what would surely be suicide, or they could slink back into the night. And one by one the rebels receded, with their crab claws cluttering their mouths and their amulets thudding against their hearts. They returned to their homes and plantations, and they tried to sleep, wondering what would happen the next day.

Within ten days every one of the organizers had been arrested except one. Gullah Jack was nowhere to be seen. People thought maybe there was some truth to his claim of divine protection. Maybe he was floating in the ether even then, laughing down at the failed rebellion. But on the eleventh day Gullah Jack's friend and coconspirator Denmark Vesey was sentenced to death. The judges spat out the words of his sentence, accusing Vesey of having corrupted the word of God to his own ends. Vesey, a humble carpenter who had accepted the mantle of leadership among Charleston's black rebels, was told: "You have committed the grossest impiety, in attempting to pervert the sacred words of God into a sanction for crimes of the blackest hue. It is evident that you are totally insensible of the divine influence of that Gospel, 'all whose paths are peace.' " And Vesey was sentenced to die by the very men who touted the gospel of peace. In the only written evidence introduced in the trial, Abraham Poyas asserted Vesey's heartfelt belief that God stood by the oppressed and would not leave the liberators alone in their time of need. But in their time of greatest need, just before they were hanged from the gallows, the hand of God was nowhere to be seen. Vesey and five others were executed on July 2, 1822, and with them died their fervent hope for divine intervention.

Maybe Gullah Jack came back down to earth to witness his friend's execution. Or maybe he had never disappeared at all. But on July 5, three days after Vesey's death and just one day after Charleston's annual Independence Day celebration, the city guard apprehended Jack Pritchard, who had been hiding in the city. During his trial it became clear some blacks were terrified of his spiritual powers. George Vanderhorst did not want to testify against Pritchard because he was afraid of his "conjuration." He pleaded with the court to keep his identity a secret, and after his testimony he asked to be moved from the region, fearing that those loyal to Gullah Jack would harm him. Harry Haig and Julius Forrest admitted they were members of Jack's Gullah company but claimed their participation was not of their own free will. They argued that Pritchard had put a spell on them that forced them to go along with his wishes, and it wasn't until Pritchard was arrested that his power over them began to wane.

Justice was swift for Jack; after only one day of testimony, he was sentenced to death. The court was eager to proclaim the impotence of Gullah Jack's religion in their sentence: "Your altars and your Gods have sunk together in the dust. The airy spectres, conjured by you, have been chased away by the superior light of Truth, and you stand exposed, the miserable and deluded victim of offended Justice." The next day, July 12, 1822, he was hanged from the same gallows where Vesey had swung just a week before. Pritchard, for his part, never cared what the whites thought of his religion. His only fear was that Vesey's spirit would torment his soul for having failed to rescue Denmark from the gallows.

While George Wilson's betrayal served as the initial crack in the rebellion's armor, it was the testimony of some of Vesey's confidants—

CLASS No. 1.

Comprises those prisoners who were found guilty and executed.

Prisoners Names.	Owners' Names.	Time of Commit.	How Disposed of.
Peter	James Poyas	June 18	Hanged on Tuesday the 2d July, 1822, on Blake's lands, near Charleston.
Ned	Gov. T. Bennett,	do.	
Rolla	do.	do	
Batteau	do.	do.	
Denmark Vesey	A free black man	22	
Jessy	Thos. Blackwood	23	
John	Elias Horry	July 5	Do. on the Lines near Ch.; Friday July 12.
Gullah Jack	Paul Pritchard	do.	
Mingo	Wm. Harth	June 21	Hanged on the Lines near Charleston, on Friday, 26th July.
Lot	Forrester	27	
Joe	P. L. Jore	July 6	
Julius	Thos. Forrest	8	
Tom	Mrs. Russell	10	
Smart	Robt. Anderson	do.	
John	John Robertson	11	
Robert	do.	do.	
Adam	do.	do.	
Polydore	Mrs. Faber	do.	
Bacchus	Benj. Hammet	do.	
Dick	Wm. Sims	13	
Pharaoh	— Thompson	do.	
Jemmy	Mrs. Clement	18	
Mauidore	Mordecai Cohen	19	
Dean	— Mitchell	do.	
Jack	Mrs. Purcell	12	
Bellisle	Est. of Jos. Yates	18	
Naphur	do.	do.	
Adam	do.	do.	
Jacob	John S. Glen	16	
Charles	John Billings	18	
Jack	N. McNeill	22	
Cæsar	Miss Smith	do.	
Jacob Stagg	Jacob Lankester	23	Do. Tues. July 30.
Tom	Wm. M. Scott	24	
William	Mrs. Garner	Aug. 2	Do. Friday, Aug. 9.

Denmark Vesey and others were found guilty and hanged. White authorities made a point of setting examples of as many "conspirators" as possible.

Monday Gell, for example—that dealt the literal deathblow to the movement's leaders. One can only imagine what might have provoked Gell and others to bear witness against their fellow blacks. The easy answer, of course, is self-interest. Things were hard enough as it was for free blacks like William Penceel; he certainly did not need the dreams of overly ambitious slaves interfering with his freedom. His testimony earned him $1,000 from the state legislature and an exemption from the state tax on free blacks. He used $700 to buy an enslaved woman and her two children, quite possibly his family.

As for Monday Gell, his very life was at stake. Maybe his only motive for cooperating was saving his own neck, and his confession did just that. The court recommended that Gell and the other informants be set free without any punishment, maintaining that they "regard it to be politic that the Negroes should know that even their principal advisors and ring leaders cannot be confided in, and under the temptation of exemption from capital punishment they will betray the common cause."

With half of the convicted defendants holding membership in the African Church, George Wilson may have thought that his cooperation with the authorities would protect the congregation from future harassment. After all, his reputation as a Christian and a class leader was on the line, and committing a sin of omission that brought harm to any of God's children was just as deadly as committing murder. The white South Carolina legislature thought that he deserved more than a clear conscience, so they awarded him his freedom and a fifty-dollar annuity. If self-preservation was their motive, most of the informants achieved their goal. At the same time, one cannot discount the role religion might have played in their decisions to come forward. George Wilson was deeply conflicted about his role in the eventual execution of thirty-five blacks. He believed it was God's will that lives be saved and probably did not anticipate the execution of so many members of his community. In the end the only lives he helped to save were white. And perhaps Gell told the truth when he said he confessed "as a man who knows he is about to die," and his only motive in making his testimony was to clear his conscience and meet God in peace. In the final analysis, though, not even Wilson's life was spared, for his decision to provide information against people he considered friends drove him to the brink of sanity, and his only way out was to take his own life.

The collapse of the most extensive plot to overthrow whites in the slaveholding South, George Wilson's Judas-like betrayal of the cause, and the execution of thirty-five black men left Charleston's black survivors wondering where God's hand was in it all. Where was God when Vesey and the others swung by the neck? Where was God when black people were being beaten bloody in the workhouse? If Gullah Jack could use his power to get others to join the movement, why couldn't he save himself? But Vesey and the other martyrs knew that their deaths would not be in vain. For beneath the rubble that was once a glorious plan to liberate God's children from captivity rested the seeds of "righteous discontent," nurtured by the spiritual striving of a people who refused to believe their God would withdraw his loving hand.

> *Father, I stretch my hand to thee,*
> *no other help I know.*
> *If thou withdraw thyself from me,*
> *Whither shall I go?*

2

The Prince

Let there arise out of you a band of people
Inviting to all that is good, Enjoining what is right,
And forbidding what is wrong.

HOLY QURAN (3:104)

One can only speculate on how Denmark Vesey's life would have turned out had he made different choices. What if he hadn't taken four years to plan the rebellion and it had gotten off the ground before being discovered? What if he had been able to enlist the military support of Haiti in his quest to rid South Carolina of slavery? How might his life have been different if he had left America to make a new start in Liberia?

By 1820 a small but significant number of blacks gave up on the possibility of enjoying full equality and liberty in the United States and sought refuge in the Promised Land of Africa. For those born on the African continent, it was a repatriation of sorts, although few actually found their way back to their ancestral homelands. When they initially departed from Africa, they had belonged to the tribes of the Ewe, Fulani, Asante, Ibo, Yoruba, Mandingo, and Hausa. They spoke different

In 1829 David Walker, born free in 1785 in Wilmington, North Carolina, published *Appeal to the Colored Citizens of the World,* a poignantly written abolitionist text that argued for universal emancipation of Africans.

languages, maintained different cultures, worshiped different gods. Yet situations largely beyond their control thrust these ethnic groups and many others into a common caldron of suffering and servitude that the world had not experienced before.

Time and circumstance forged them into a new people—American, but not quite; African, but not quite. Some left Africa giving praise to Allah or the gods of their clan and returned practicing the religion of white people. Yet though they prayed to the same God as those who claimed blacks were inferior to whites, many African Americans saw themselves as guardians of the faith. Albert J. Raboteau asserts, "The passion of Jesus, the suffering servant, spoke deeply to the slaves, the sorrow and pain of their life resonating with his. For if Jesus came as the suffering servant, the slave certainly resembled him more than the master." They saw in Christianity possibilities for freedom of spirit and body. Teaching subservience and acceptance of slavery they took to be distortions of the teachings of Jesus. Regardless of whether they called on Jesus, Allah, or some other deity, they believed that their gods were stronger, more powerful, indeed more benevolent than the God of their captors. It was this faith that convinced them that they could cross the Great Sea and return home one day. Their faith was not only sustenance in a new world of uncertainty, it was also their most potent weapon as they struggled to maintain belief in the face of life's trials.

Most repatriated African Americans settled along the coast of Sierra Leone and Liberia, colonies established for freed men and women of African descent. Founded in 1792 and 1820 by Great Britain and the United States, respectively, these countries quickly became symbols of the possibilities of African American self-governance and social progress. Moreover, the opportunity for building new churches in the African colonies was not lost on black and white Christians in the United

States. Daniel Coker, one of the most influential figures in the independent black church movement, even decided to set sail for the land of his ancestors to spread the gospel among native Africans and those returning from the Americas. A man of sharp intellect, he penned one of the first antislavery tracts by a person of African descent, *Dialogue Between a Virginian and an African Minister.* Coker was also the founding pastor of Bethel Church in Baltimore and the first man to be elected bishop of the AME Church. He traveled to Africa as a missionary because he was frustrated by his inability to unite Baltimore's black Methodists in a single independent church. In the winter of 1820, as the ship *Elizabeth* carried black emigrants from New York to Liberia, Coker organized a society of Methodists on the open sea. Once he and the other passengers landed, they continued to worship together and planted the first seeds of African Methodism in Liberia and Sierra Leone.

Had Denmark Vesey chosen to cast his lot in the land of his forebears, he might have ended up a member of Coker's worshiping community. He might even have been standing at the shore one day in 1829, when a ship carrying Ibrahima Abdul Rahman docked. A tall, slim man whose keen features and dazzlingly ebony skin were as impressive to whites as the dignity with which he lived his life, Abdul Rahman was returning to the African continent in bittersweet victory. Although he reached his goal of taking his wife out of the United States with him, Ibrahima's American-born children remained in Mississippi. And though he managed to land on African soil, he remained a long distance from the region in which he was reared. Yet he was pleased to be in a land where service to Allah was more than a distant memory or a quaint curiosity. He could unabashedly serve the God of his choosing and feel spiritually whole again. For him life had come full circle, all in accordance with the will of Allah.

Freed Africans in 1880 from the Mount Olivet Baptist Church in Arkansas preparing for a departure to Liberia. In the periods preceding and following slavery significant numbers of African Americans supported emigration to Africa.

Ashmun Street in Monrovia, featuring a Methodist Episcopal church and a Roman Catholic mission house, was home to many prominent settlers in Liberia.

It took Ibrahima Abdul Rahman nearly four decades to make it home to Africa, but in that time he never relinquished his belief in the power of Allah, nor did he waiver in his commitment to the principles of Islam. His experiences in bondage mirror those of countless other Muslims not mentioned by name in historical records. Their stories form the corpus of an American jihad—not in the traditional understanding of a "holy war" waged against nonbelievers but as an inner struggle for self-improvement and moral purity that is manifested outwardly through society's transformation. Ibrahima Abdul Rahman's American jihad began less than one year after the United States Constitutional Convention. By the time he arrived in chains in New Orleans on

Ibrahima Abdul Rahman as he appeared in New York in 1828, in a sketch by Henry Inman.

June 9, 1788, the slave trade in North America was well into its second century.

Born in 1762 in Timbuktu, Abdul Rahman was a member of the Fulani people, an ethnic group from West Africa's interior. They were cattle herders known for their ferocity in battle and fervent commitment to their God, Allah. Like most Muslim boys in the region, Abdul Rahman began his Quranic education around age seven, learning to read and write in Arabic as well as Pulaar, his native language, which was transcribed into the Arabic alphabet. He learned to recite his prayers in Arabic, sitting on a sheepskin mat, and he wore a leather amulet around his neck with a Quranic passage for protection tucked inside. Long before he developed any notion of being an African, Abdul Rahman was a Muslim who felt the strongest connection to those on the continent who were servants of Allah.

Islam had been entrenched in sub-Saharan Africa since the fourteenth century, although it arrived on the continent much earlier. Traders carried the faith along well-established trade routes. Contact with strangers in distant regions afforded African Muslims the opportunity to speak about their faith. Even though often they continued to practice their indigenous religions, some African rulers recognized the socioeconomic advantage of being part of a religion as mobile and international as Islam. At least they recognized the importance of the protection the Islamic kingdoms offered. By the eighteenth century the primary method of conversion shifted from commerce to war. Despite this fact, Muslim Africans and their animist neighbors lived side by side and, for the most part, tolerated one another's beliefs. This was particularly true in urban areas, where religious traditions melded frequently. It was in the cities that Muslims had the most contact with non-Muslim cultures. However, in times of conflict Islam provided a convenient justification for African Muslims to overpower their non-Muslim adversaries. Armed with the Quranic mandate to fight on behalf of Islam when necessary and the assurance that "God will certainly aid those who aid His [cause]," warriors for the faith expanded the reach of Islamic empires in West Africa.

Timbo, in present-day Guinea, was the largest city in Futa Jallon, the home of Abdul Rahman and his ancestors. The city was a home base for the herders when they returned from the pastures, and it served as an intellectual hub of the kingdom. Literacy was very important in West Africa, and Futa Jallon was no exception—the region could boast thousands of schools. In 1774 twelve-year-old Ibrahima

THE SPREAD OF ISLAM IN AFRICA

Islam came to Africa from the north and east. After Muslims invaded Egypt, they went in three directions: from the Red Sea to the eastern coastal lines, up the Nile valley to the Sudan, and west to the Maghreb. (These events occurred up to the eleventh century.) By the twelfth century, Islam overshadowed Christianity in North Africa, and by the fifteenth century it had reduced the highly influential and popular Egyptian Christian community to a small minority. From 1050 to 1750 Islam spread south across the Sahara and along the Nile.

Until the nineteenth century, Islam in East Africa remained on the periphery, the coastline. East Africa had a direct line of communication via the Arabian peninsula. And the new class of merchants and landowners contributed to the religion's growth in East Africa.

Islam reached South Africa starting around 1798. Its rapid spread there is attributed to slaves migrating from Malaya and the islands of the Indian Ocean.

left his family in Futa Jallon for further study at Macina in present-day Mali, about 1,000 miles east of Timbo. He also spent some time under the tutelage of Muslim clerics at Timbuktu, twelve days' travel past Macina. In Timbuktu he was exposed to Arabic translations of the Pentateuch, the New Testament, and advanced Islamic texts along with the basic tenets of Judaism and Christianity.

He displayed a passion for learning, particularly material related to Islam. Unfortunately, his studies were cut short in 1779, when Abdul Rahman returned home to join his father's army. The Fulani were locked in a fierce battle with animists in the region. Like many communities throughout Africa, the Fulani occasionally warred with their neighbors in long-standing feuds over territory. The periodic call to arms in defense of the kingdom solidified both physical and attitudinal boundaries between young Ibrahima and his neighbors. He was Fulani, they were not. To him, they were barbaric. He was Muslim, they were not. For him the lines were as clear as the African sun was hot, and while the Fulani had limited social interaction with other tribes anyway, young Ibrahima was taught that Islam required a certain degree of social separation to maintain the purity of the faith.

Prisoners of war between the tribes were used for agricultural labor and were traded to the Europeans as slaves bound for the Americas. As the Fulani acquired more and more slaves in the late eighteenth century, they had more time to devote to the study of religion. The Quran was translated into their native language, and they erected numerous mosques and Quranic schools.

While it is true that slavery existed in Africa long before the arrival of Europeans, there were important distinctions between the two systems. Unlike slaves in the transatlantic trade, enslaved Africans on the continent had certain rights and privileges. They owned land; in some cases they were fully adopted as members of their owners' clans; and they often had an impact on the culture of their owners. They married, raised families, and their status was not necessarily transferred to the next generation. For Muslim Africans, the slaving enterprise was theoretically regulated by Islamic legal theory. Technically, Muslims were prohibited from enslaving members of the faith, although violations of this prohibition are well documented. According to Islamic law as well, the child of a free person and a slave was born free.

As the years progressed and the trade with Europeans increased, slavery was no longer simply a natural by-product of war—it became the primary reason for war. Africans traded prisoners of war for guns in attempts to subdue other groups, which merely intensified the fighting and increased the demand for guns. This cycle resulted in a rapid growth of the supply of African labor, matching the seemingly insatiable demand for human chattel in the New World.

In 1781 the Fulani had an unexpected visitor who would play an important role in Abdul Rahman's future. John Coates Cox was an Irish physician who had been separated from his party near the West African coast and was spotted wandering about by a Fulani scout unit. Emaciated, wounded, and gravely ill, he was taken to Timbo, where he immediately became a local curiosity. No one in town had ever seen a white person before, and Cox's novelty necessitated a guard to keep at bay the onlookers intrigued by his pale skin. Nonetheless, he was treated with the same respect granted other visitors and nursed back to health. Nearly two decades later Cox recalled that the Fulani "treated me as kindly as my own parents."

Ibrahima spent every spare moment with Cox, learning all he could about the world beyond the Great Sea. Cox probably taught him some very basic English and some of the elementary precepts of medicine. The doctor enjoyed the time he spent with the young prince and his people so much that he contemplated remaining with them. Nevertheless, once healed of his jungle fevers Cox was overcome by homesickness. When he was ready to return to the coast and wait for a European vessel to carry him home, Ibrahima's father sent an armed escort of fifteen men along with Cox to ensure his safety, as well as "gold to pay his passage home." The cultural exchange between Cox and Ibrahima was mutual and sincere, and both men benefited from their months together.

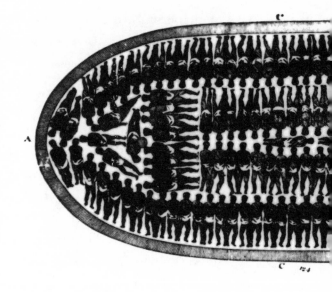

Abdul Rahman's next contact with Europeans was not as positive. In 1788 he was captured by a neighboring group during battle. After a weeklong journey down the Gambia River to the coast, he was sold to European traders for a few guns and powder, two bottles of rum, and eight hands of tobacco. For the next six weeks he lay in the belly of the slave ship *Africa* on its way to the Caribbean. Death made a daily visit to the vessel as its human cargo wilted under the inhumanity of their new living conditions. The men, women, and children on board came from various parts of the African continent and held varied and often unflattering opinions of one another. Yet the will of Allah had thrust them together in the bowels of a floating tomb.

Abdul Rahman took comfort from the fact that other Muslims were on board and from his knowledge that *al-Wali*, God the Protector, would never leave him defenseless. He and the other Muslims on the ship did their best to perform their Islamic obligations, but conditions made it impossible to fulfill their religious requirements strictly. Ritual purification before prayer was an impossibility, but that did not deter them from praying when possible. They might have prayed fewer than the required five times daily, but they prayed nonetheless. When brought on deck in the morning, they would face the rising sun and open with the words uttered by Muslims for over a millennium: *Bismillahir rah-*

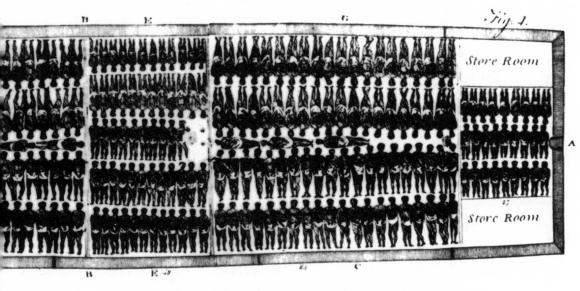

A typical layout of a slave ship plying its trade through the Middle Passage across the Atlantic. Africans were crammed into spaces eighteen inches high for a ten-week voyage.

manir rahim, "In the name of God, the Merciful, the Compassionate." Only by faith in Allah would Abdul Rahman and his fellow Muslims make it through the Middle Passage.

A full month and a half later *Africa* docked in the Caribbean at the island of Dominica, and Abdul Rahman was still among the living. He had endured the most difficult trial of his life, but his travels and travails were far from over. After a brief stay in Dominica he was placed on a ship headed toward New Orleans, one more week's travel to the northwest. Once in the Crescent City, he was shipped up the Mississippi River and finally disembarked four weeks later in the Natchez District of Louisiana in summer 1788. After several months and 6,000 miles of travel, his faith in God and his people remained unshaken. To give up the struggle meant to surrender hope, and Abdul Rahman refused to believe that enslavement was God's ultimate intention for his life. If he could only get word to his family that he was still alive, they would spare no expense in securing his return. Using one of the plantation slaves as an interpreter, he told Thomas Foster, his new owner, of his royal lineage. But Foster was unimpressed. Most likely in jest he responded to the tale of grandeur by naming his recently acquired property Prince.

It is quite possible that whoever interpreted for Abdul Rahman had knowledge of his family background. Abdul Rahman was probably not the only enslaved African on the plantation from Futa Jallon, nor was he the sole practitioner of Islam. In fact, many of the planters in the Natchez District preferred slaves who were Mandingoes, as they called African Muslims. Some ethnic groups, in particular the Ibo, were thought to be more prone to despair and suicide as a way to escape enslavement and return home. But whites in Natchez generally believed enslaved Muslims to be more intelligent and better suited for leadership than slaves from other tribes. One scholar notes, "In Islam, there is no separation between religion and the secular life.... So that everything that you do during the course of a day reflects the religion—whether it's what you eat, what you drink, the way you dress, how you behave, etc., Islam tells you exactly what to do during the whole course of your day and the whole course of your life."

Since Islam had strict guidelines for daily living, white slave owners observed that enslaved Muslims did not drink alcohol, exhibited a high degree of discipline, seemed to pick up the English language more easily than others, and did not engage in social dancing and other activities in the slave quarters. As a result, they were selected disproportionately by the planters to act as drivers and overseers. Planters especially familiar with the Fulani thought them to be intelligent but unsuited for hard labor. Abdul Rahman certainly was unaccustomed to farm labor and thought it beneath him, but the more he resisted the more he was beaten.

As the weeks passed, Abdul Rahman continued to rebel against his servitude. Determined to break him, Foster became increasingly violent. Prince eventually decided to steal away into the woods, armed only with his faith in Allah and his determination to be free. His amulet likely was gone, but he knew from memory the verse it contained. Reciting it in Arabic brought great comfort in the midst of his uncertain situation, and as the nights grew dark, damp, and cold, Abdul Rahman never forgot that Allah was there with him.

God, there is no god but He, the Living, the Everlasting.
Slumber seizes Him not, neither sleep;
to Him belongs all that is in the heavens and the earth.
Who is there that shall intercede with Him save by His leave?
He knows what lies before them and what is after them, and they
comprehend not anything of His knowledge save such as He wills.
His Throne comprises the Heavens and the earth; the preserving of
them oppresses him not; He is the All-high, the All-glorious.

The *Ayat al-Kursi,* or Verse of the Throne, is a Quranic passage used for centuries as assurance of Allah's protection against evil forces. And while Abdul Rahman had been in situations before where he needed such assurances, the swampy woods were utterly foreign and terrifying to him.

After weeks on the run, he dealt with the limitations of his new freedom. There was neither manna from heaven to nourish him nor daily bread delivered in the beaks of ravens. Wild boar were plentiful, but eating swine was against his religion. He held on to the possibility that he could make his way back to the river that led to the Great Sea, but even if, by the grace of Allah, he made it to the coast, he could not get to Futa Jallon on his own. Certainly it would be only a matter of time before slave catchers discovered his hiding place. Islam did not permit him to take his own life, so his only option would be to return to the plantation. After all, if Allah was indeed the author of all things, his capture and removal from his native land must have been a part of Allah's will. At the same time, Abdul Rahman never gave up believing that Allah wanted him to be free.

When he returned to the Foster plantation, he was greeted with severe beatings. Abdul Rahman decided to make the best of a bad situation. He did not relinquish his struggle to be free, but he chose to resist in new ways. Initially he did so by exploiting white generalizations about him. If his owner believed that he was different from the other slaves because of his religion's moral code, perhaps he could take advantage of that belief to make his life a bit easier. He developed a trusting relationship with his owner; Foster attested that he had "never known him intoxicated, (he makes no use of ardent spirits)— never detected him in dishonesty or falsehood," nor was he ever "guilty of a mean action." This is, most likely, why Abdul Rahman was given choice jobs and allowed to continue to practice Islam. Though it was difficult for him to hold on to his religion without the support of a community of believers, he used every means available to preserve his faith. He wrote Arabic letters in the sand during breaks from work in order to maintain his connection with the language and the religion.

As time went by Abdul Rahman's owner grew to admire his Muslim slave more and more. Watching Abdul Rahman, Foster observed, "Though born and raised in affluence, he has submitted to his fate without a murmur, and has been an industrious and faithful servant." Perhaps he thought it was Islam that made Ibrahima appear so content with his station in life, but nothing could have been further from the truth. What Foster interpreted as faithfulness to him was to Abdul

Rahman faithfulness to Allah. By maintaining a commitment to his religious beliefs, Ibrahima waged his own private war against his captors.

In some places one finds a more public manifestation of the struggle of enslaved Muslims. In enslaved communities that were heavily Islamic, like the Bahia region of Brazil, Muslims were at the helm of uprisings and other rebellious activity. However, for Muslims like Abdul Rahman, living in isolated communities on relatively small plantations, such behavior was not as common.

Resigning himself to the fact that he likely would never see his family in Africa again, Abdul Rahman started a new family in Mississippi. On Christmas 1794 he married Isabella, one of two enslaved women on the Foster plantation. Isabella had been born in America, although historians have no information about her life before her marriage. We do know that she did not share her husband's religious convictions. A devout Christian since her conversion to the Baptist faith in 1797, she raised their nine children as Christians. Unlike some other denominations, Baptists offered a certain degree of acceptance to blacks in the South at the close of the eighteenth century. Because of the emphasis placed on what one historian described as "democratic inclusiveness" in Baptist congregations, enslaved blacks were "received as equal members of congregations along with whites, baptized at the same time as whites, called Brother and Sister, and given the 'right hand of fellowship' equally with whites."

As black Baptist numbers increased, however, democratic inclusiveness was replaced by Negro pews and "Nigger Heaven" balconies. Despite the blatant racism in the churches, Abdul Rahman believed there was much in Christianity that was comparable to Islam. The problem, he reasoned, was the way Christians selectively followed the precepts of their religion. Years later, when asked about his impressions of Christianity, he responded, "I tell you the Testament very good law;

you no follow it; you no pray often enough, you greedy after money." Without becoming a Christian, he attended services at the Presbyterian church where his master held membership, and from 1816 until his emancipation twelve years later, Abdul Rahman accompanied his family regularly to services at the local Baptist church.

Even while attending Christian worship, Abdul Rahman made attempts to share his Islamic beliefs with his children. But there is no indication that any of them subscribed to such tenets. There are a number of reasons that one finds evidence of few second-generation Muslims in the antebellum United States. The lack of Arabic texts, particularly the Quran, was a major impediment to the transmission of Islam. Abdul Rahman had been enslaved for nearly thirty-five years before he gained access to any sort of printed matter in Arabic. And even though most Muslims memorize significant sections of their holy text, their ability to recall such passages quickly faded without ways to reinforce retention. Furthermore, the absence of Quranic schools virtually eliminated any possibility of passing on Islamic law and its history of interpretation. Unlike American Sunday schools, Quranic school was an all-day affair every day. Obviously, the work schedule of enslaved Africans did not permit such an intensive educational experience. Additionally, plantations in the United States were, with a few important exceptions, relatively small compared with those in the Caribbean and South America. This meant that there were relatively few enslaved African Muslims on any one plantation. Without the kind of community support necessary to sustain and transmit religious knowledge, Abdul Rahman stood little chance of rearing his children as Muslims.

The inability of Islam to adapt its faith requirements and practices also proved a barrier to its transplant in North America. While minor adjustments in Islamic practices to the working and living conditions in which enslaved Muslims found themselves were made, the core

FIVE PILLARS OF ISLAM

1. FAITH (*Shahada*)

The confession of Allah as God and of Muhammad as his holy messenger is the rock upon which Islam rests. Muslims declare "that there is no God but the one God and Muhammad is His prophet." The recitation of *shahada* is a sign of one's faith and belief in the Islamic tradition.

2. PRAYER (*Salat*)

Muslims are required to perform *salat* (prayer) five times per day: at dawn, noon, midafternoon, sunset, and nightfall. These prayers function as one's direct link to Allah. Muslims may pray in any clean area at work, at home, or in a mosque.

3. ALMSGIVING (*Zakat*)

Zakat, a kind of taxation, is one of the most significant pillars of Islam. Indeed, Islam teaches that all material goods belong to God; hence, goods and property are actually gifts lent by God. It is the duty of Muslims to share their property and to give away portions of their income to help the poor. This philanthropy constitutes a form of purification for the giver and in turn multiplies the giver's blessings.

4. FASTING (*Sawm*)

During the ninth month of the Muslim calendar, known as Ramadan, all followers fast for thirty days from food, drink, and sexual activities, beginning at sunrise and going to sunset. Muslims see fasting as a means of purifying the self and reinforcing their commitment to Allah. In addition, the discipline of abstinence stands as a reminder of Allah's numerous blessings in their lives. For health-related reasons, Muslims may fast in other ways that illumine their faith and obedience. The thirty-day fast is concluded with a celebration known as Idd al Fitr.

5. PILGRIMAGE (*Hajj*)

All healthy and financially able Muslims are strongly encouraged to make a pilgrimage to Mecca at least once in their lives. This pilgrimage should occur during the twelfth month of the Islamic year. During the *hajj* Muslims wear simple clothing, minimizing class barriers. The rituals performed in the pilgrimage retell and honor the story of the origins of Islam and enable Muslims to recommit themselves to the pillars of their faith.

beliefs of African Muslims in North America were never Christianized. In fact, Muslim rituals and practices were incorporated into the religious practices of enslaved blacks. Ironically, it was the quest to protect the purity of Islam among enslaved blacks that, in the end, led to its virtual extinction in the North American antebellum period.

While Ibrahima Abdul Rahman was isolated from any sort of community of devout Muslims, he always depended on his faith in Allah. He used his distinctiveness as a Muslim to distance himself from other enslaved blacks, even going so far as to proclaim to a white newspaper reporter "explicitly, and with an air of pride, that not a drop of Negro blood runs in his veins." Indeed, the fact that his hair did not seem to kink into tight coils and his facial features were unlike those labeled by Europeans as Negroid only strengthened white Natchez's opinion of Prince as different from, and superior to, other slaves.

Islam seemed to have a de-Africanizing effect in the minds of white plantation owners, and Abdul Rahman was determined to get as much mileage out of white ignorance as possible. He intimated to interested whites on more than one occasion that he believed Negroes to be "infinitely below" his own people. He also never missed an opportunity to tell all who would listen the stories of his royal lineage, military battles fought and won, and his family's immense wealth and power among the Fulani people. White observers often described him as confident and dignified, whereas blacks saw him as arrogant with a tinge of moral superiority.

Abdul Rahman's peerless conduct earned him several important benefits. He and his wife were given the privilege of working a plot of land for themselves, the surplus produce of which they sold in town for cash. Access to hard currency, however little, afforded him higher social standing among the enslaved population. He also met a number of influential people who took an interest in his story. One day in 1807, while at the market selling vegetables with Sambo, another enslaved African, Abdul Rahman saw a man who looked very familiar. Although life's hardships had taken a toll on the stranger, Abdul Rahman was positive he had met the fellow before. "I observed him carefully, and knew him," said Prince, "but he did not know me." After a brief conversation about his potatoes, the fellow asked the forty-five-year-old Prince, "Boy, where did you come from?" Once he answered, the white man recognized him as well, and this seemingly chance encounter broke open a dam of emotions that flooded Abdul Rahman's soul.

It was Dr. Cox, the gentleman Ibrahima and his people had befriended in Africa. Without regard for local social mores, the two men embraced. They talked about how their lives had changed since they last met, and Prince gave a full account of his capture, transport, and experience in bondage. Cox's heart went out to Ibrahima, for he knew what it was like to be stranded in a strange land and to fear that the gentle touch of loved ones would be never felt again. He knew what it was like to live with foreign languages, foreign customs, and foreign gods. He knew that despite the passage of time Prince longed for the land of his forebears. Most important, he knew that before him stood the opportunity to show his gratitude for what the Fulani people had done for him.

Over the next few months Cox relentlessly worked to persuade Thomas Foster to sell Prince to him, but Foster refused even to set a price. Even after Cox's death his son, also a physician in the Natchez area, continued the effort to purchase Ibrahima's freedom. But Foster wouldn't consider it. Nonetheless, the verification of Abdul Rahman's storied past made him something of a celebrity about town and enhanced his value with Foster. Everyone knew now what was only suspected before—that there was something special about this man they called Prince.

In 1818, after thirty years of toil in the blazing Mississippi sun, fifty-six-year-old Ibrahima Abdul Rahman was delivered from field labor. The remainder of his life in captivity was spent in relative comfort, with the majority of his workday unsupervised. Selling vegetables at the market in town became his new responsibility. Perhaps his advanced age made him less useful as a field hand, or maybe Foster valued more the social capital accrued from flaunting Prince—a trustworthy Muslim of royal lineage who had faithfully served his master for three decades. Around 1820 Andrew Marschalk, editor of the *Mississippi State Gazette*, made Abdul Rahman's acquaintance, and over the next five or so years the two men talked occasionally. Prince began to frequent Marschalk's print shop.

His stories of African life fascinated Marschalk, and in their long conversations the men shared their views on religion. Because Abdul Rahman seemed unique in so many ways, Marschalk imagined a unique origin for the Prince. He incorrectly assumed that Prince was Moroccan, a non-Negro nationality that Europeans believed exhibited a higher degree of civilization than people from sub-Saharan Africa. Actually, the only thing that connected Abdul Rahman to Morocco was Islam, but Marschalk repeatedly told others that the man they called Prince was no ordinary Negro but a Moor. Abdul Rahman apparently did little to correct the error, for he knew that such a myth would hold Marschalk's interest. Cyrus Griffin, another Natchez journalist, knew more about Africa than Marschalk, and he immediately recognized that Futa Jallon was nowhere near Morocco. Ethically, he felt it necessary to correct the mistake, but he did not want to jeopardize Abdul Rahman's best shot at freedom.

On one of his visits to the print shop, Prince informed Marschalk that he wanted to write a letter to Futa Jallon but had no knowledge of how to get it there. Marschalk promised him that if Prince wrote the letter, he would make sure it was safely sent to Africa. Abdul Rahman refused to write, however, and gave no apparent reason for his hesitation. Perhaps he was not sure if Marschalk could be trusted, and was concerned about Foster learning of his grab at freedom, or maybe he thought whites would lose interest in him once they learned he was not really from Timbuktu. Whatever the reason, he would not write the letter, despite Marschalk's constant urging.

Then one day in 1826, Prince strolled into Marschalk's office ready to write the letter. No one knows for sure what brought about his change of heart. Marschalk did not seem to care about the reasons. He knew that Prince's story was fascinating, and hoped that publicizing it would increase the chances of Abdul Rahman gaining his freedom.

Although Abdul Rahman had lost most of his Arabic, certain passages of the Quran still came to him without much effort. Instead of crafting a plea for ransom, he wrote in Arabic one of the few things he could remember. The letter did not survive, so scholars are uncertain of its content, but whenever he was asked to write something in Arabic, Abdul Rahman almost always penned the Fatiha.

> In the name of God, the Most Gracious, Most Merciful. Praise be to God, the Cherisher and the Sustainer of the Worlds; Most Gracious, Most Merciful; Master of the Days of Judgment. Thee do we worship, and Thine aid we seek. Show us the straight way, the way of those on whom Thou hast bestowed Thy Grace, those whose portion is not wrath, and who go not astray. Amen.

The Fatiha is the opening chapter of the Quran and is an integral part of both canonical and spontaneous prayers. Recited by Muslims throughout the world, it is one of about twelve passages that Muslims must memorize in Arabic. Maybe the Fatiha was all Abdul Rahman could write. But he doubtless knew that when the letter was sent to Morocco, it would convey to its recipient that Abdul Rahman was a Muslim—a man committed to the same God in whom Moroccans placed their faith.

News of Ibrahima Abdul Rahman's life quickly became more than just a local-interest story in the Mississippi dailies. The U.S. State Department got involved in Abdul Rahman's quest for freedom, because Marschalk led them to believe Prince was from Morocco. Perhaps it was simply a case of his ignorance of African geography, or perhaps he knew that such a claim would attract more attention than saying Abdul Rahman was a prince of Futa Jallon. Regardless of Marschalk's motives, the fact that the U.S. government believed Abdul Rahman to be Moroccan played a crucial role in their decision to become involved. They thought that intervening on his behalf would be a good diplomatic move if they could exchange him for a guarantee of better treatment of shipwrecked Americans off the northern coast of Africa.

Abdul Rahman's letter reached Tangier, Morocco, five months later, on March 14, 1827. Once the Moroccans showed interest in Abdul Rahman, the only thing left to do was convince Thomas Foster to grant his freedom. Foster agreed to let Prince go only if his freedom would be enjoyed outside the United States.

On February 22, 1828, Ibrahima Abdul Rahman, prince of the Fulani people, was again a free man, yet his sojourn in a strange land was far from complete. How could he leave his wife of nearly thirty-four years and their children behind? Africa was not their homeland, but he believed they deserved the same chance at freedom he did. After weeks of intense negotiations, Foster agreed to release Isabella to Marschalk for $293. Neither Abdul Rahman nor Marschalk could afford $293, but through door-to-door subscriptions it took them only twenty-four hours to raise the money. By the middle of March, Isabella was free for the first time in her life. While Ibrahima and his wife were at liberty to make the long trip to Africa, their nine children and their families remained in bondage. They both knew that getting Foster to agree to manumit them for any price would be extremely difficult. They had to settle for hoping that one day their clan could be reunited.

Ibrahima and Isabella were ready to embark, and Andrew Marschalk took care of all the arrangements. Secretary of State Henry Clay authorized Marschalk to spend $200 for the trip to Washington, D.C. He took $70 of the money and purchased an outfit—some would say costume—for Prince. It included a white turban with a blue crescent, a blue coat with shiny yellow buttons, white pants gathered at the ankles, and "Moroccan style" yellow boots. It was a brilliant fund-raising strategy that played on white fascination with the exotic aspects of the "Dark Continent." Marschalk booked the couple on a steamboat set to depart from New Orleans on a twelve-day journey to Cincinnati via the Mississippi and Ohio Rivers, with stops in Memphis and Louisville. In Cincinnati, Abdul Rahman was able to raise enough money to cover his wife's expenses, since the State Department had made no provisions for her. At the end of April they left Cincinnati headed up the Ohio River to Wheeling, West Virginia; then they traveled overland to Baltimore.

Once in the city Ibrahima and Isabella found themselves in an extremely awkward situation. Maryland had the largest number of free blacks—53,000—in any of the slave states. Baltimore's black community of nearly 19,000 accounted for 25 percent of the city's total population, and over 75 percent of that number were free blacks. As a result, there was a strong movement among whites to rid the state of its free black denizens. Abdul Rahman's visit was supported in part by the Maryland Colonization Society, one of the groups hoping to send free blacks to either Liberia or Haiti. While he was paraded about town in attempts to raise money to free his children, the city's black community was not nearly as inviting as he would have liked. In their estimation Abdul Rahman was being used by the Colonization movement as a poster child for repatriation. That may have been the intent, but Abdul Rahman was willing to accept help from any individual or group that showed interest in his cause. It became clear that most of the people in Baltimore thought he was from Timbuktu. Moreover, they assumed that, since his father was dead, he was the heir apparent on his way home to claim the throne. Abdul Rahman neglected to correct his hosts, choosing instead to play on their curiosity about the place still unspoiled by Europeans.

Ibrahima and Isabella still wanted to free their children from bondage, but they were uncertain how to get the money to pay Thomas Foster. Ibrahima arranged to meet Secretary of State Henry Clay, who was visiting Baltimore at the time, and Clay urged him to travel at once to Washington, D.C., to meet the president, John Quincy Adams. Ibrahima left his wife behind to reduce the expense of the trip. The two men had a pleasant meeting, although Abdul Rahman's diplomatic usefulness plummeted once Adams learned that he was not Moroccan. The prince informed Adams that he wanted to go to Liberia, then home to Futa Jallon. Also, he expressed his desire to remain in the States until his children were freed.

Adams was deeply disappointed that he could not use Abdul Rahman for diplomatic purposes with Morocco and decided that the politics of the matter would not permit him to intervene on behalf of the prince's children. He informed Abdul Rahman that the American Colonization Society did not purchase individuals but merely assisted blacks already free in getting to Africa. Abdul Rahman was urged by supporters in Baltimore to visit Philadelphia to raise money for his wife's passage and his children's purchase. Leaving his wife in Baltimore, he spent the next six weeks in Philadelphia but was able to raise only $350. Discouraged but not defeated, he used some of the proceeds to travel further, visiting several towns and cities before ending up in Boston.

Unlike the black community in Baltimore, Boston's blacks seemed unanimous in their enthusiasm for Abdul Rahman's visit and his quest to make it back to Africa. They held a parade and banquet in his honor and took up a collection for the purchase of his children. At the banquet virtually every man in the community rose to offer a special toast in praise of their guest. After the sixth salutation David Walker, known for his passionate advocacy of immediate emancipation, rose to add a few words. "Our worthy Guest, who was by the African's natural enemies torn from his country, religion, and friends, and in the very midst of Christians, doomed to perpetual though unlawful bondage may God enable him to obtain so much of the reward of his labor, as may purchase the freedom of his offspring."

The son of a free woman and a slave father, Walker was born in 1785 in Wilmington, North Carolina. Despite his father's being a slave, Walker was considered emancipated at birth because of his mother's status. Yet his black skin was his social marker; his legal status carried little, if any, protection from racism. Walker spent his early years traveling around the South, observing day-to-day encounters with racism and injustice. "Having traveled over a considerable portion of the United States," he wrote, "the result of my observations has warranted the full and unshaken conviction, that we [colored people of the United States] are the most degraded, wretched, and abject set of beings that ever lived since the world began." He believed Abdul Rahman, however, was a clear example of what blacks could achieve if they settled for nothing less than their freedom.

After settling in Boston in 1826, Walker began systematically working toward the eradication of racism. He opened a clothing boutique and quickly connected himself with prominent black Bostonians involved in antislavery efforts. He joined a local Methodist church, became an active participant in the Massachusetts General Colored Association, and was the Boston agent for *Freedom's Journal*, the first black newspaper published in North America. In just a short time, the six-foot-tall, dark-skinned, statuesque North Carolinian became an important fixture in Boston politics in general and the black antislavery movement in particular. His most important accomplishment, the 1829 publication of *Appeal to the Colored Citizens of the World*, represented a voice from the margins and advocated for both individual and collective activism against slavery and the American Colonization Society. Published less than a year after Abdul Rahman's Boston visit, the essay was distributed through the postal system with the help of black and white seamen, who took copies to ports south of the Mason-Dixon line.

The *Appeal* blasted the institution of slavery and leveled a scathing critique of the colonization scheme, but these passages did not compare with Walker's harsh criticism of white American Christians. He pointed out that other religions, like Judaism and Islam, offered protection and acceptance of those who converted to their faith, but American Christianity allowed slavery. Walker asked, "Have not the Americans the Bible in their hands? Do they believe it? Surely they do not. See how they treat us in open violation of the Bible!!" If white Christians truly believed the words of the Bible, he thought, they would have no choice but to admit the sin of slavery.

Walker and the other black Bostonians assembled in Abdul Rahman's honor recognized the symbolism of his quest for freedom and his desire to return home. Dressed in the costume Marschalk had bought for him, the prince bore witness to the resiliency of the African spirit. The fact that he remained a Muslim stood as a critique of white Christian America and its desire to break the will of enslaved Africans, and even though most of the people in the room were concerned about the far-reaching implications of colonization, they all supported Abdul Rahman's yearning to free his family and reunite with them on African soil.

But there were still some rather important obstacles to overcome. The American Colonization Society thought it might be able to show Abdul Rahman the error of Islam and persuade him to spread the message of Christianity through West Africa. Although Abdul Rahman respected Christianity, in particular the New Testament, he had no interest in becoming a Christian. Nonetheless, he was willing to do

All persons who are acquainted with history, and particularly the Bible, who are not blinded by the God of this world, and are not actuated solely by avarice—who are able to lay aside prejudice long enough to view, candidly and impartially, things as they were, are, and probably will be, who are willing to admit that God made man to serve him *alone,* and that man should have no other Lord or Lords but himself—that God Almighty is the sole *proprietor* or *master* of the WHOLE human family, and will not on any consideration admit of a colleague, being unwilling to divide his glory with another.—And who can dispense with prejudice long enough to admit that we are men, notwithstanding our *improminent noses* and *woolly* heads, and believe that we feel for our fathers, mothers, wives and children as well as they do for theirs. —From the Preamble

Are we MEN!!—I ask you, O my brethren! Are we MEN? Did our creator make us to be slaves to dust and ashes like ourselves? Are they not dying worms as well as we? Have they not to make their appearance before the tribunal of heaven, to answer for the deeds done in the body, as well as we? Have we any other master but Jesus Christ alone? Is he not their master as well as ours?—What right then, have we to obey and call any other master, but Himself? How we could be so *submissive* to a gang of men, whom we cannot tell whether they are as *good* as ourselves or not, I never could conceive. However, this is shut up with the Lord and we cannot precisely tell—but I declare, we judge men by their works. —From Article I or say whatever was necessary to secure his family's passage to Africa. Thomas H. Gallaudet, an American known for his work with the hearing- and speech-impaired, started corresponding with the prince in an attempt to influence him spiritually. He even sent him an Arabic copy of the New Testament. Abdul Rahman wrote a letter thanking Gallaudet for the Bible and pledging his support in the effort to Christianize Africa.

> After I take this book home, I hope I shall get many to become Christians. If I find things at home in the same way I left, I think they will become Christians. When I left my country almost all the young people followed the Christian religion. Whether they continue to follow it, I know not. When I take home the two books, the Arabic Testament and Bible that you sent me, I think they will follow the Christians.

Abdul Rahman rode Christianity as far as it would take him. When asked to write the Lord's Prayer in Arabic for potential financial contributors, he would write the Fatiha instead. So even as he claimed to have converted to Christianity, he continued to hold on to his faith.

Gallaudet was not the only person interested in Abdul Rahman for what he could do for Western society. Charles Tappan, a wealthy white New York businessman whose family was deeply involved in antislavery and colonization causes, showed a great interest in the prince's plight, too. Although his concern was partly a result of his commitment to Christian philanthropy, Tappan thought Abdul Rahman might be the missing link between the United States and Timbuktu—that African commercial and cultural center that, for Westerners, represented the untapped resources of the continent. Since his father was born in Timbuktu, Tappan and others incorrectly assumed that Abdul Rahman was

A variation of the Fatiha written as the Lord's Prayer by Ibrahima Abdul Rahman in December 1828.

the prince of Timbuktu and its immediate environs. Moreover, they thought that as soon as the prince returned he would be elevated to king. Abdul Rahman allowed Tappan to hold on to his dream of African riches as long as the businessman was willing to assist in the procurement of his progeny in Mississippi. In reality, Abdul Rahman probably never expected to see Timbuktu again.

Finally, on February 7, 1829, almost one year after Thomas Foster had freed him, Ibrahima Abdul Rahman and his wife were ready to make the voyage to Africa. They had raised nearly $3,400 to cover expenses and help pay for the manumission of their children, but they could no longer afford to wait for them to be free. Along with 152 other blacks seeking a new life in the African sun, they boarded the ship *Harriet* in Norfolk, Virginia. When they arrived in Liberia nearly forty days later, Abdul Rahman was physically and emotionally spent from the journey. Nevertheless, he did not neglect to give Allah thanks and praises. He agreed to serve as a bridge between Washington and Timbo in the hope that his assistance might result in some form of governmental intervention on behalf of his offspring. But the rainy season prevented the sixty-seven-year-old

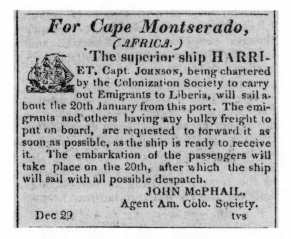

For Cape Montserado,
(AFRICA.)
The superior ship HARRI-
ET, Capt. JOHNSON, being chartered
by the Colonization Society to carry
out Emigrants to Liberia, will sail a-
bout the 20th January from this port. The emigrants and others having any bulky freight to
put on board, are requested to forward it as
soon as possible, as the ship is ready to receive
it. The embarkation of the passengers will
take place on the 20th, after which the ship
will sail with all possible despatch.
              JOHN McPHAIL,
          Agent Am. Colo. Society.
Dec 29   tvs

A newspaper advertisement announcing a voyage to Africa sponsored by the American Colonization Society, which helped steer the emigration movement to Africa.

from making the fifteen-day trip from Liberia to Timbo, and on July 6, 1829, Ibrahima Abdul Rahman died.

Abdul Rahman's quest for freedom yielded mixed results. Although he lived only five months after his return to Africa, he was able to die a free man on the continent of his birth. While he never got to see his children again, a year after his death some of his descendants did join his widow in Liberia. Freedom cost him a large emotional toll, but Abdul Rahman was one of the few enslaved Muslims in the United States who maintained his commitment to Islam and died holding on to the beliefs of his ancestors. Throughout his life he served Allah well, and he was rewarded with a glimpse of things to come—a renewed relationship between the descendants of those who were snatched from Africa decades earlier and their spiritual kin on the continent.

I sell the Shadow to Support the Substance.
SOJOURNER TRUTH.

Speak to My Heart

He said to me: O mortal, stand up on your feet,
and I will speak with you. And when he spoke to me
a spirit entered into me and set me on my feet; and I heard
him speaking to me. He said to me, Mortal, I am sending
you to the people of Israel, to a nation of rebels who have
rebelled against me; they and their ancestors have
transgressed against me to this very day....
You shall speak my words to them, whether they hear
or refuse to hear; for they are a rebellious house.

—EZEKIEL 2:1–3, 7

Isabella watched in 1826 as the morning sun burned through the blanket of fog over the Hudson River Valley, signaling a new day. There would be no more hours filled with labors of sorrow for her. Today was the day she planned to walk away from the Dumont homestead after nearly thirty years of enslavement. She did not run. She did not steal away in the middle of the night. She waited for morning and "walked away by day-light" with the confidence that justice and God were on her side. The tall black woman bundled her few clothes

Sojourner Truth, born in upstate New York around 1797, a profound
preacher and abolitionist who spent some forty years traveling North America
speaking out against slavery and injustices against women.

and some food in a large, square, plaid cloth. She had no apparent destination. With the bundle over her shoulder and a baby on her hip, Isabella was risking all with only God by her side, believing in the One who had brought her through all her previous trials and tribulations.

At six feet, she looked most men in the eye, and her sinewy arms could match most men's strength. But now she needed another kind of strength. It would take great spiritual courage to get away from her master and escape other white slave owners as she sought the light of liberty. Her hair was unkempt, her clothes torn and tattered, and her feet bare, but she was finally free from the harsh labor doled out by her owner, John J. Dumont. She was also free from the humiliating verbal abuse of Dumont's wife, Elizabeth. Certainly Dumont would track her. The odds favored him. He would be on horse; she was walking. But she didn't care. In fact such a confrontation was exactly what she wanted, so she could tell him why she'd left in the manner she did.

The spirit of the Lord led her to the home of the Van Wageners, a Dutch Reformed couple. Isabella had met them at a rally where the Van Wageners spoke against slavery. They said owning human beings was a violation of God's law. When she reached their home, just five miles away, Isaac Van Wagener stood outside to talk with her. She spoke with such pain that he agreed to lodge Isabella. But John Dumont arrived later that morning to reclaim his property. Isabella stood to her full six feet and informed him she was no longer his slave, "because you had promised me a year of my time." Emancipation Day was set in New York State for July 4, 1827, but Dumont had told Isabella she could have her freedom a year earlier. Although this early release was couched as a reward for past labors, Dumont and many other slave owners in the state thought it might be best to free some slaves ahead of schedule rather than risk a year filled with fake illnesses and inactivity. However, Isabella had cut her hand during spring planting, and the cut had become severely infected, limiting her ability to work. Dumont decided that her loss in output meant she had to remain in bondage until Emancipation Day.

Hurt hand and all, Isabella knew she still did the work of two laborers, so she considered their arrangement intact. Oddly, it was fairness not slavery that was the issue in this confrontation. Isabella was an extremely fair-minded and honest person, always completing her assigned chores to the best of her abilities. As one who worked in both the field and the house, she had access to all the Dumonts owned, yet she never took anything that did not belong to her. "The Lord only knows how many times I let my children go hungry," she said, "rather than take secretly the bread I liked not to ask for." This was not simply because she wanted to make her owner happy but because, she said, "it made me true to my God." So when Dumont reneged on their deal, she felt justified in claiming the time she had coming to her.

Standing outside the Van Wageners' farmhouse, Isabella displayed a fierce manner that she had rarely shown at Dumont's home. She dared him to grab her, and he backed down. He was about to get back on his horse but not before he found a way to claim victory over this woman he had owned. None of her five children was ever a part of their verbal contract. It was spite and a desire to keep Isabella in her place that fueled Dumont's sudden interest in Sophia. If he could not take Isabella, he would seize her child Sophia. He ripped the infant from her blanket as she lay on the Van Wageners' porch. Isabella pulled back, trying to hold her daughter. As the two tugged at the child, Van Wagener rushed over. He offered to buy Isabella and the child from Dumont, even though this offer violated his own antislavery impulse. Then God's inspiration appeared, the mystical "ram in the bush" that rewarded true believers in the Bible. Van Wagener offered to purchase a year's worth of Isabella's services, basically leasing her and the children until emancipation. This was a turning point for Isabella.

Isabella Bomefree was born in 1797 to James and Elizabeth Bome-free, slaves in Ulster County, New York. Their owner was Johannes Hardenbergh, head of one of the best-known Dutch families in the state. Unlike that of the Deep South, Ulster County's landscape was not suitable for large plantation agriculture. The ground was rocky, with steep hills protruding above the Hudson River. The majority of the families in the area were Dutch, so Isabella grew up speaking only that language. It was an isolated existence for enslaved blacks, with typically only about five slaves per slave-owning family. In 1800 less than 5 percent of New York State's total population was enslaved, compared with roughly 40 percent for Mississippi.

While the institution of Northern slavery varied from its Southern counterpart in significant ways, it maintained one of Southern slavery's hallmarks—the emotional and psychological brutality of domination. As commodities enslaved blacks lived with the constant threat of being sold, wreaking havoc on families. Enslaved parents were at their most desperate when they tried to protect their children from being sold to strangers. Ma Bet (as Isabella called her mother) knew too well the pains of losing a child, having had seven of her nine children sold away from her. Only Isabella and her younger brother remained, and although she could not ensure they would always be with her, Ma Bet did what she could to instill in them the belief that they would never be alone. "My children, there is a God, who hears and sees you," she told them. "He lives in the sky…and when you are beaten, or cruelly treated, or fall into any trouble, you must ask help of him, and he will always hear and help you." Isabella never forgot what her mother told

THE PARTING "Buy us too."

A slave woman cries out for mercy as she
witnesses her husband being taken away by a slaveholder.

her, and over the years God became not only her shelter in times of trouble but also a dear friend.

Who was this God to whom Ma Bet entrusted her children's lives? Was he the God she had brought with her across the waters from Africa? Was he the God of Christianity, first person of the Holy Trinity? Most likely he was a bit of both. Ma Bet was influenced by the Dutch Calvinism taught in the Reformed tradition throughout the Hudson River Valley. By 1794 Dutch Protestants in the United States had seceded completely from their parent body in Holland and formed the Reformed Church in America. "The primacy of God's power in human life is at the heart of the preaching of the [Reformed] church," notes one historian, and it was God who directed the paths of believers. For the Dutch Reformed, God was powerful and yet distant. One approached God with both reverence and deference. A central part of Ma Bet's religious instruction to her children was the Lord's Prayer. It was a way to let God know one's needs without appearing to force or cajole God into action. In Dutch Ma Bet taught Isabella that all she needed to do was utter the words of that prayer and the totality of her needs—from daily meals to protection from evil—would be addressed.

The God of Ma Bet's African past was much more personal; this God was concerned with her daily activities and struggles. She knew it was necessary for her children to have God to lean on at any minute. So in addition to the Lord's Prayer, she taught them to talk with God in a personal fashion. As one professor later commented:

Mama Bett wanted her daughter and her son to see God as someone who was very close. This was a personal friend. God was certainly supernatural. God was all-powerful. God was omniscient. But God was also very close. God could be as close as the ancestors. God could be as close as her mother. So God was a warm personal friend, someone whom Bell could talk to as she would talk to her mother. When her mother wasn't there, then Bell could talk to God. This was certainly something that Bell took with her for the rest of her life, that God was there for her, and she had but to call Him and talk to Him.

This meant that while God at times appeared remote, he was in reality no more than a prayer away.

Ma Bet taught Isabella always to go outside under the "canopy of heaven" and address God in the midst of his created order. She believed that prayers in the great outdoors had a better chance of being heard and answered, since there the lines of communication were unfettered. And Isabella believed prayer should be spoken out loud, just as one talks with a close friend. "She had no idea God had any knowledge of her thoughts, save what she told him; or heard her prayers, unless they were spoken audibly," said one biographer and friend, "and consequently, she could not pray unless she had time and opportunity to go by herself, where she could talk to God without being overheard." Isabella wasn't talking to herself. She fully expected God to answer in an equally clear manner.

Isabella listened for God's voice when she struggled with her decision to leave the Dumont farm. As her mother had taught her, she went to the brush arbor she had constructed on a tiny spit of land in the middle of a shallow stream. The rush of the water provided just enough noise to prevent her conversation with God from being overheard. There she and God talked about the unfairness of Dumont's action and how she would respond. Once it was clear that leaving was her only recourse, she asked God to help her formulate a plan. She could not leave when people were up and about for fear she would be seen. But leaving at night was not an option, for she had no idea

The earliest-known portrait of Sojourner Truth, done around 1850.

where she was going. All of a sudden God told her what she was to do. If she left at dawn, she would not have to fear the dark and could still get to wherever she was going before Dumont missed her. "That's a good thought!" she told God. "Thank you God for *that* thought!" And, like the prophet Ezekiel, she moved to and fro guided only by the spirit of God.

Isabella's spirituality took on an added dimension in 1827. As she turned thirty she experienced God in a new way. It was early summer and, among the region's Dutch African population, that meant it was time for Pinkster. Traditionally a Dutch Reformed observance of Pentecost, Pinkster had a primarily secular meaning to blacks in the Hudson River Valley. It was a time for socializing, drinking, and dancing. Isabella had been a hardworking, hard-drinking, and even wild young woman. She remembered how much fun she'd had during Pinkster at the Dumonts' and longed for a chance to visit with friends and family there. She appreciated all the Van Wageners had done for her, but their strict Calvinist piety was far less exciting than life at the Dumont farm. Isabella was not particularly devout, although she certainly considered herself honest, just, and kind. She came to God in times of trouble, but when all was well she had less use for him. Life at the Van Wageners' was so nice that there was little room for God.

One day Isabella got the feeling that her former owner would come to visit her. It was odd, but he came by the Van Wageners' farm

REBECCA COX JACKSON (1795–1871)

While there were a handful of antebellum black women who were recognized as ministers of the gospel, most women who felt called to preach did not receive support or official recognition. The sexism of the day used scriptural interpretation to argue for the subjugation of women, and leadership positions in American Protestant churches were virtually closed to them. One antebellum black preacher who has gained significant notoriety since the discovery of her spiritual autobiography in 1981 is Rebecca Cox Jackson.

Born free in Horntown, Pennsylvania, Rebecca was raised by her mother and grandmother. In 1830 she married Samuel S. Jackson, and the couple moved in with her eldest brother, Joseph Cox, who was an AME preacher and tanner in Philadelphia. She reared her brother's four children while working at home as a seamstress.

After a powerful thunderstorm in 1830, Jackson experienced a religious awakening that transformed her life. She claimed it was revealed to her that celibacy was necessary if Christians were to lead a holy life. She started preaching this message in small groups that would meet in women's homes and immediately drew the ire of the all-male AME leadership and clergy. As one might expect, her position caused problems in her own home as well and eventually led to her separation from the men in her life.

Jackson spent the 1830s and part of the 1840s as an itinerant minister, preaching the message of the Holiness movement and freedom from the pleasures of the flesh. She later joined the United Society of Believers in Christ's Second Coming, commonly known as the Shakers, in Watervliet, New York. The group's emphasis on personal spiritual experience and celibacy appealed to her and satisfied religious yearnings that went unfulfilled by the AME Church. Jackson remained a member of that community until 1851, when she and Rebecca Perot, her younger protégée, left for Philadelphia. They believed the society was not doing enough to take the message of Mother Ann Lee, its founder, to black communities. They established a community of black Shakers there that taught the power of Holiness and celibacy mostly to black women. Jackson died in 1871, but not before inspiring a new generation of black Shakers to turn inward in order to experience all God had to offer.

regularly to see her. They argued, but they also laughed and had long conversations. And this time she made up her mind that she would return with him to his place to celebrate Pinkster. Sure enough, he did come, and when she got ready to leave with her child, something came over her. It was a sense of being part of something outside of herself that was looking at her desire for drinking and dancing. Although she could not see this thing, she could feel its power around her like a mighty wind and she could not escape its condemnation. She believed it was God's way of telling her that a return to the Dumonts' and the sinful ways of the past would be a fatal turn away from his goodness.

Isabella could not move for a few moments, and her breathing became labored. "Oh God, I did not know you were so big," she responded once she realized that the presence she felt was God. She tried to go back in the house and work in the hope that her chores might relieve her mind and body of this burden, but it was not possible. As she began to think that she needed someone or something to stand between her and God to protect and guide her, another presence revealed himself. After a short while she recognized the presence—it was Jesus. Immediately she felt the warmth and comfort of a loving friend who wanted only what was best for her. It was Jesus who eased her fear of God, who had paid the ultimate price for her sins, and who would forevermore walk by her side.

Throughout her life Isabella maintained this dichotomy between God and Jesus, and as the years wore on she saw Jesus more and more as her intermediary. "I did not see him to be God," she said, "else, how could he stand between me and God? I saw him as a friend, standing between me and God, through whom, love flowed as from a fountain." Isabella believed she knew Jesus personally, and she was quite surprised to learn that others knew him just as well. It did not seem to occur to her that he could be a best friend to so many people at the same time. How could he be for others exactly what he was for her? In the years following her conversion experience, she even became jealous when she discovered that others had an equally intimate knowledge of Jesus. But as she matured in her Christian walk, she understood that it was unreasonable to have Jesus all to herself.

In 1828 Isabella moved with Sophia and her teenage son, Peter, to New York City. Life in the city was a totally new experience for her, one that provided both support and challenges to her faith. She found that New York's seedy side was fertile ground for spreading the word of God. Isabella took to the task almost immediately. She worked in the Manhattan home of a wealthy white family and in her spare time joined her employer and another white woman to offer salvation in the city's

Broadway where it crosses Canal Street in New York City, circa 1830.
Sojourner Truth lived on Canal Street on and off throughout the 1830s.

many taverns and brothels. The three ladies became part of an interracial reform movement in the city that was interested in initiating a
spiritual revival among the unchurched classes. Her first extemporaneous speaking and preaching was done in New York as a part of
this work.

Isabella initially joined the Methodist church on John Street. After
a year in the city, however, she became part of a black Christian movement and joined Zion Church on Church Street. This church was
composed of blacks who were former members of the John Street con-

gregation. Although the John Street Church had become over 40 percent black by the early 1790s, its members of African descent routinely were told to sit in the back. They were also told to wait until whites were finished before approaching the altar for Holy Communion, and they held no positions of church leadership. In 1796 a group led by William Miller and Peter Williams withdrew from the congregation and founded the African Chapel in a shop owned by Miller. By the time Isabella arrived in New York, the group had changed its name to Zion Church and had joined with another black Methodist congregation to form the African Methodist Episcopal Zion Church, a black Methodist denomination independent of white control and separate from Richard Allen's AME Church.

Even though she considered Zion her spiritual home, Isabella did not confine her religious work to the congregation. She spent most of her time with reformers working among the city's outcasts. But while Isabella found receptive audiences among the city's poor, she was having serious problems keeping her own son on the straight and narrow path. Isabella had worked to buy Peter after Dumont had taken out his anger at her by selling the boy to a Southern slave owner. But when she brought him to New York, she found that she could not control the intelligent young man's curiosity about every aspect of city life. He was always on the verge of trouble.

Isabella did all she could to instill her faith in the children and keep them from the sinful life so easily available in the city. But she spent a lot of time away from home and Peter was drifting away. "I took them to the religious meetings," she said. "I talked to, and prayed for and with them; when they did wrong, I scolded at and whipped them." What more was a mother to do? Isabella finally decided not to help Peter the next time he got into something, in the hope that an experience in jail would do him some good. In the end her tough love paid off, and after a few nights in jail Peter decided to flee the snares of the city by becoming a sailor. Isabella missed her son dearly when he left in the summer of 1839, but she knew that God would take care of him and that he was better off away from the negative influences of New York. In order to save him she had to let him go.

She continued to think about and pray for her son, but Peter's absence gave Isabella the opportunity to focus even more on her religious activities. Like many people in New York, she was hungry for interaction with the Holy Spirit and sampled widely from the varieties of Christian expressions available in the city. The Second Great Awakening had engulfed a large portion of New York State along the Erie Canal, from west of Albany through Rochester and Buffalo. This area became known as the Burned Over District because everyone in it seemed to be ablaze with the Holy Spirit. New York City also experienced the effects of the state's revivalist impulses, as many of the new prophets of God chose to locate where they felt the need was the greatest. On May 5, 1832, Prophet Matthias visited the home where Isabella was working. This man looked very odd to her with his long hair and beard. At first she thought he might be Jesus in the flesh. Though he assured her he was not, he said he was in the city to do God's work. Since the master of the house was not at home, they had a conversation about religion in which he told her that he was a Jew.

A mostly white Christian crowd attending a camp revival during the Second Great Awakening in the early nineteenth century.

Although she was a bit confused, some of his ideas made sense to her, and she said she "felt as if God had sent him to set up the kingdom." She made up her mind then to labor with him in God's vineyard and learn all she could from this strange yet seemingly powerful visionary.

The one thing Isabella discovered while under the spiritual guidance of Prophet Matthias was the difficulty of living a faithful life in a sin-filled city. She told the prophet that the Holy Spirit made it clear to her that God had work in store for her elsewhere—particularly in New England. Although she had never been to New England and had no friends there, she told one of her female acquaintances that "the Spirit calls me there, and I must go." In June 1843, after fifteen years, she

MEMOIRS

OF

MATTHIAS THE PROPHET,

WITH A FULL EXPOSURE OF

HIS ATROCIOUS IMPOSITIONS,

AND OF THE

DEGRADING DELUSIONS OF HIS FOLLOWERS.

PRICE THREE CENTS.

WRITTEN FOR THE NEW YORK SUN.

NEW YORK.

OFFICE OF THE SUN.........222 WILLIAM STREET.

1835.

Title page of the memoirs of the Prophet Matthias,
billed as *In the Kingdom of Matthias*.

left New York the same way she'd arrived—with only a few changes of clothing and enough food to last her a couple of days. She felt compelled to take on a new name, one that fit her new vision of herself. She became Sojourner Truth and left New York City, preaching God's word wherever she found a receptive ear.

Itinerant ministry was a rather common practice in the early nineteenth century, and religious women took full advantage of it to be free from male authority. While Christian denominations were not ordaining women to be priests or pastors, religious women carved out a space in which they could fulfill their duty to God and work around church doctrine and polity. Naturally those movements on the fringe of American Christianity were often most receptive to women as preachers. After leaving New York, Sojourner Truth embarked on a storied career that led her throughout the Northeast.

She was an extemporaneous preacher, never preparing but simply allowing God to speak through her. The historian Nell Painter notes, "Sojourner Truth said that she didn't prepare ahead of time. And she was always curious to see what she was going to say. She felt divinely inspired, whether she was talking about God, or religion, or women's rights, or anti-slavery. She felt that God spoke through her." Sojourner Truth could not read, but she did not allow her illiteracy to prevent her from studying the Bible. She memorized significant sections of the text, particularly passages from the four Gospels. And whenever she needed to consult the scriptures she enlisted children to read to her. Adults could not, in her opinion, be trusted to relay the words unedited and without comment, and she did not want anyone else's views to stand between her own reasoning and the Truth of the Word.

By the time she left New York City, Sojourner Truth was a full-fledged Millerite. Following the teachings of William Miller, a self-taught

Mrs. Juliann Tillman, an early-nineteenth-century
preacher in the African Methodist Episcopal Church.

Baptist minister whom his supporters believed had calculated the exact date of the Second Coming of Christ, Sojourner Truth also thought that the end of the world was near, and that Jesus would soon return to judge the living and the dead. Miller claimed that sometime between March 21, 1843, and March 21, 1844, the "sanctuary shall be cleansed" with holy fire. As a Millerite, Sojourner Truth was convinced there was no time to spare. Souls needed to be saved, which meant the Word of God had to be preached. She had to tell all who would listen about Christ's imminent return. She walked all over Long Island and through the Connecticut River Valley, preaching wherever she found a receptive ear and relying on the Providence of God and the kindness of strangers for food and shelter. For the most part, she counted on word of mouth and a network of Millerites.

The winter of 1843 found her in western Massachusetts, and the prospect of a cold winter with no place to live was daunting, but she knew that God would continue to provide for her. It turned out that she had more than enough offers to stay in utopian communities. Influenced by the religious vision of Robert Owen and his communitarian project in New Harmony, Indiana, experimental communities seeking to live at one with nature and humanity abounded in the Northeast. Brook Farm in West Roxbury, near Boston, was one of several communities that offered Sojourner Truth a place to stay, as did the United Society of Believers in Christ's Second Coming, popularly known as Shakers for their enthusiasm in worship. In the end, she spent the winter at the Northampton Association of Education and Industry in Northampton, Massachusetts.

Sojourner Truth's choice of Northampton was significant for several reasons. The association was part of a much larger communitarian movement. Some of the groups were explicitly Christian, while others offered new models for improving American secular society. In the 1840s fifty-nine new communities were established in the United States. More-over, these groups' growth generated new connections to social reform movements nationwide. Northampton, for example, was a hotbed of abolitionism, and many of the association's members were involved in the effort to outlaw slavery. While the group had only a handful of African American members, Isabella did not feel isolated in North-ampton because of her race. It was there that she came to know David Ruggles, a radical black abolitionist from New York.

Ruggles, born in Lyme, Connecticut, in 1810, was widely known in abolitionist circles for his work with the New York Committee of Vigi-lance. He had helped over six hundred people, including Frederick Douglass, escape slavery and settle in the North. Ruggles also champi-oned the rights of free blacks in New York City, as well as safeguarded fugitives against recapture.

Sojourner Truth had been acquainted with Ruggles's work in New York City, but by the time she arrived in Northampton he was nearly blind and in very poor health. Nonetheless, his passion for freedom and justice had not diminished, and the members of the community drew strength from his stories of battles for liberty's sake. Sojourner Truth had not really been active in the abolitionist movement before her arrival in Northampton; to that point all of her energy had been focused on moral reform movements in and around New York. But in Northampton she became increasingly involved with the abolitionists and started speaking on the subject just as often as she did on the return of Christ.

As March 1843 and March 1844 came and went, cynicism began to set in about William Miller's predictions of the Second Coming of Christ. He made one final adjustment, claiming October 22, 1844, was the exact date that everything prophesied in the Book of Daniel would come true. But soon the great expectation of the rapture turned into

Sojourner Truth delivering one of her prophetic messages
before a diverse crowd of women and men.

what many called the Great Disappointment. It was enough to send Millerites searching for truth outside the Christian faith, or to doubt the existence of God altogether. Although Sojourner Truth was embarrassed by her commitment to Miller's predictions, she did not question the reality of God and Jesus. They had been too important in her life for her to doubt their care and concern for her.

Sojourner Truth joined the antislavery speaking circuit and became well known for her quick wit and fiery tongue. Even though most white abolitionists accepted her with open arms, she often sensed hostility from the crowds she addressed. In 1858, while she was giving an antislavery lecture in Indiana, the proslavery Democrats in the region came out to heckle. A local doctor accused her of being a man in disguise. He demanded that she go backstage and expose her breasts to some of the women present to verify her sex. She replied that she had nursed plenty of white children during her years of enslavement, often to the exclusion of her own, and some of them had grown up to be men despite having nursed on a black woman's breast. In fact, in her opinion they were far better men than any present that evening. She then calmly pulled out her breast and asked if any of the men wanted to suck.

She had made her point: that the Lord was her strength and salvation, and she need not fear the physical and psychological intimidation of any man. She was indeed a woman, a black woman who spoke the truth as God revealed it to her. And in her mind the message was clear—slavery was the sin of America, and God would punish his people if they did not repent and turn from their wicked ways. Rather than continuing to berate audiences about their need for repentance and salvation, she turned her energies toward America's social ills, specifically slavery and inequality between men and women.

Barracks at the Freedmen Village in Arlington Heights, Virginia,
where Sojourner Truth lived and taught in the 1860s,
now the site of Arlington National Cemetery.

From the time she left Northampton in 1846 until her death in 1883, Sojourner Truth fought for freedom and equality under the banner of God's Golden Rule—Love thy neighbor as thyself. In addition to her antislavery work, Sojourner Truth was an activist for women's rights. In her mind the two causes were parts of the same problem: that some of God's creatures believed God favored them over others. She knew this belief to be false, and never tired of telling people that God showed no favoritism. But her blackness often was a liability in the women's movement. Men who opposed women's rights tried to force white women to distance themselves from black women and their concerns, and Sojourner Truth's broken English and illiteracy could be seen as evidence of women's innate physical and intellectual inferiority. Moreover, Christian men occasionally claimed male superiority because of Jesus. Clearly, they reasoned, even God preferred maleness.

All these issues were present at the 1851 Women's Rights Convention in Akron, Ohio, where Sojourner Truth was scheduled to speak.

Some of the white women believed allowing her to address the crowd would turn the occasion into an "abolitionist affair," overshadowing the validity of their own claims for equality. Nonetheless, Sojourner Truth addressed the crowd and directly confronted the men who opposed women's rights.

> Dat man ober dar say dat women needs to be helped into carriages, and lifted ober ditches, and to have de best place every whar. Nobody eber help me into carriages, or ober mud puddles, or give me any best place... and ar'n't I a woman? Look at me! Look at my arm! I have plowed, and planted, and gathered into barns, and no man could head me—and ar'n't I a woman? I could work as much and eat as much as a man (when I could get it), and bear de lash as well—and ar'n't I a woman? I have born thirteen children and 'em mos' all sold off into slavery, and when I cried out with a mother's grief, none but Jesus heard—and ar'n't I a woman?

Sojourner Truth's life as Isabella disappeared as she grew to be an American legend. She lost touch with her roots in upstate New York. Her eldest daughter, Diana, once wrote to her that Dumont, her former master, had moved west to be with his sons in his old age. After a few letters she never heard from Peter again. But her life was full. She became even more spiritual in her sixties and seventies, taking part in séances and speaking of a God that roared like an open fire across all denominations, all states, and every color of man or woman.

She lectured widely and kept the little money she was paid. With her savings she was able to buy property in Michigan in 1856, and after the Civil War she petitioned Congress to give land in Kansas and other western states to the former slaves. Congress never acted, but by the time Sojourner Truth died in 1883, at the age of eighty-six, the nation had heard the voice of the Almighty coming from the mouth of this former slave woman who called them to righteous treatment of all citizens as children of God.

"God Is a Negro"

By the rivers of Babylon,
There, we sat down, yea, we wept
When we remembered Zion.
We hung our harps
Upon the willows in the midst of it.
For there those who carried us away asked of us a song,
And those who plundered us requested mirth,
Saying, "Sing us one of the songs of Zion!"
How shall we sing the Lord's song in a foreign land?

—PSALMS 137:1−4

As Bishop Henry McNeal Turner surveyed the crowd at Atlanta's Friendship Baptist Church, he planted his feet wide, put his hands on his hips, and stood like a triumphant Christian soldier. For years he had battled criticism from both blacks and whites who denounced his call for black people to leave the white church. Even within his beloved African Methodist Episcopal Church, many believed that Turner's appeal to racial pride in the cause of creating a free-standing all-black church did little more than stir up enmity between

The Reverend Henry McNeal Turner, an outspoken bishop in the African Methodist Episcopal Church. Turner was a strong proponent of emigration to Africa and regularly criticized racial injustices against African Americans.

the races. And his vision of a separate black church was particularly dangerous at a time when blacks and whites struggled to coexist in the political turmoil that shook the defeated South after the Civil War. He was accused of being a race baiter and a tool of white racists in their efforts to get black people out of white institutions. Turner saw himself as a man of God offering a hand to Southern black people, former slaves who had never had any church or school to call their own.

Despite frequent, loud, and public condemnations of his ideas, black Baptists invited Turner to address their convention on a warm September afternoon in 1895. The black Baptists were contemplating forming their own denomination: the National Baptist Convention. The convention had the potential to create the most politically powerful organization of black people in the nation. The invitation to Turner was a challenge: could he convince the delegates to take the risk of forming a separate black religious organization? Turner stood before the thousands gathered as the voice charged to sound the clarion call. He began with the idea of black people uniting in the name of God scared more than white Christians—it scared black people, too. Turner had to ease the tension before he could rally the convention. Not every black Baptist in the room believed separation to be a good idea, especially since it would mean losing support from white Baptists in the North. Moreover, some thought that the formation of the National Baptist Convention was contrary to the cause of Christ, further dividing God's children instead of bringing them together as one pure and holy church. But Turner envisioned the power of black unity under the banner of Christ. This power was not available to former slaves in politics or through their meager pocketbooks. In his deep Southern accent he began to preach, then sing, to cry, then pray, as he told the convention that it was time for black people to claim their place as God's chosen people, called by his Son to create a miracle in the world.

Bishop Turner pounded the pulpit as he said that white Christians had made an idol of skin color when they insisted on the whiteness of God; he argued that if God had any color at all it must be black. Why? he asked. Now smiling, his voice roaring, Bishop Turner said black people had to see God as black if they saw God in themselves. He said it was whites who had taught blacks by way of enslaving their minds to see God's image and power in white skin.

Turner's speech that day had people on their feet shouting *Amen*. Even those who were stunned by his statements found themselves caught in a frenzy of spiritual revelation. For the first time they imagined God as not just with them but one of them. Turner's comments not only shook the convention but made headlines throughout America's black and white Christian communities. Henry Lyman Morehouse, corresponding secretary of the white-controlled American Baptist Home Mission Society, called the speech "the race spirit gone mad." But Turner did not back down from his assertion that to a black man God must be a Negro. The bishop continued to preach that black people had a spiritual need to hear that they, too, were created in God's image and were just as important as any other race in the eyes of God. In 1898 Turner wrote in *The Voice of Missions:*

> Every race of people since time began who have attempted to describe their God by words, or by paintings, or by carvings, or by any other form or figure, have conveyed the idea that the God who made them and shaped their destinies was symbolized in themselves, and why should not the Negro believe that he resembles God as much so as other people? We do not believe there is any hope for a race of people who do not believe they look like God.

In church pulpits throughout the South, he called on black Christians, just one generation out of bondage, to understand that they needed their own institutions to reflect God as a black man. The bishop said that God was with them. He said it was God's will that they not live as second-class Christians but stand tall so that their faithfulness would be rewarded in this life.

Bishop Turner's vision of a separate black church for America's former slaves may have come from the fact that he had never been a slave. Born free in Newberry Courthouse, South Carolina, in 1834, he had the aura of a leader—a man of distinction—from his earliest days. Turner worked in the cotton fields like the other lads his age, but he worked for his parents. And he dreamed the dreams of a free man. In one dream he saw himself at the top of a mountain. There were thousands of people gathered around to hear his message. He didn't know what the message was, but he was sure that this was God's way of telling him that his life's work was to be in teaching and preaching. He kept the dream to himself for fear others might see him as arrogant and self-righteous, but he was convinced that he had glimpsed God's purpose for him.

His sense of God's hand in his life gave the boy a strange self-assurance in a state where young black people usually displayed broken spirits. Local blacks took pride in the way he walked the streets. Whites in the area noticed his quick wit, sharp mind, and desire to learn. Around 1848 two white lawyers in Abbeville, South Carolina, began to tutor him, even though it was against the law to teach a black person to read or write. But by the grace of these white strangers and the power of his own determination, Turner got an education comparable to that of any young white person in South Carolina. By the time he was eighteen his sense of being touched by God was strengthened by his unusually high level of education for a black South Carolinian.

Turner's faith was sure that God was preparing him to lead a new Israel into the land of liberty.

At the same time Turner's interest in the life of the spirit was growing with lessons from a very different source. The old slaves, almost all of them illiterate, called him into their prayer circles. They treated him as a golden child and taught him that he was filled with the spirit. Even as a small boy Turner favored the worship style of the enslaved Christians. The faith of the slaves did not try to reach God through reason but instead sought to experience the presence of God through earnest emotion. Turner's mother worshiped with the slaves, and she often testified loudly and publicly to the goodness of God despite the daily hardships of her life. She took her son to tent revivals, where he witnessed the power of God to heal broken souls. When there was a break in planting or harvesting, black people would gather to hear the Word.

African Methodists in the "worship" experience in a Philadelphia alley in 1812 or 1813. Methodists taught strict adherence to biblical moral codes.

It was at one of these camp meetings that he found his own spiritual spark. In 1851, after years of watching others consumed by God's frightful power, the seventeen-year-old Turner was slain in the Spirit. He lost control of his body; he fell to the ground, rolled around in the dirt with limbs flailing, and sobbed. The passion that overtook his body and soul that day left its mark. This was religion at its most powerful for him. He had education, but he also had soul—comfort with vivid expressions of emotion by the faithful. With his childhood ambition of being a teacher in mind, the young Henry McNeal Turner decided to devote his life to ministry in the cause of Christ.

The 1850s were a tumultuous time for America's churches. The controversy over slavery divided the church just as it did the federal government. Several of America's largest denominations split over whether slavery should be abolished. In 1844 the Methodists divided along geographic lines over the issue. So when Turner began his ministerial career in the South in 1853, the white leaders of the Methodist Episcopal Church viewed black souls as intended by God for slavery. They would not give Turner a church but licensed him as an itinerant preacher. He was not the pastor of any particular church but was responsible for the spiritual care of those black Methodists living in his preaching circuit. He quickly gained a reputation as a learned and fiery preacher. Both black and white Christians in his region took notice of his gift. While he relished the freedom of being on his own, traveling back roads, Turner also longed to stand as a leader of all Methodists. As a first step he wanted a chance to pastor a church with a congregation he could identify as his own flock. He knew, however, his chances were slim as long as he cast his lot with the white Methodists of the South.

While visiting New Orleans in 1857, Turner attended a worship service at the St. James African Methodist Episcopal Church and was so impressed by the teachings of the Reverend Willis R. Revels that he joined this fledgling church of black Methodists on the spot. "You can put me through whatever crucible your church law demands," he told Revels, "for I am free born, and think I can stand the test." After Revels inspected Turner's preaching license from the white Methodist Church, he brought him to the front of the congregation and in the tradition of the Bible extended the young man the right hand of fellowship. And he encouraged Turner to join the AME Church by standing for examination as a preacher at the next annual conference, to be held at St. Louis in August 1858.

At the conference Turner was required to preach a trial sermon. To test his knowledge of the Good Book and the depth of his faith, he was not given the scriptural text for his sermon until moments before he was called to speak. His message had to be extemporaneous. After preaching he was questioned for three hours by a board of clergy. It was clear that the examination committee wanted to be careful whom they admitted into their ranks, for a preacher intent on charting his own course and not that of the denomination might jeopardize the AME Church's progress. The church was so small and poor, its followers lacking in education, that there was no room for mavericks who might diminish their one asset—unity. Turner tried to dispel any notion that he was a loose cannon. He saw his future as a leader of black Methodists. But the older ministers kept pounding at him with questions testing his discipline as well as his godly spirit. He began to weary under the barrage of criticism—some of which was personal and attacked his education, which threatened some of the unschooled ministers.

Finally Bishop Daniel Alexander Payne, one of the older ministers, who like Turner was from South Carolina, stepped in and told the young preacher to have a seat. In the months before the conference Payne became a mentor to Turner and put a strong emphasis on learned preaching and maintaining strong moral standards. Payne more than anyone else had confirmed Turner's instinct to pull away from the white

Daniel Alexander Payne became the sixth bishop in the AME Church in 1852. Bishop Payne engineered the founding of Wilberforce University in 1856 and became its first president.

Methodists. Now he stood up for young Turner and told the other church leaders it was time for the young man to be admitted as a preacher in the AME Church: "Now brethren, you know that this young man, brother Turner, preached a more able sermon than one-half of you can deliver, for he did not know what was going to be his text till I gave it to him as he was entering the pulpit. Yet I regard his effort as highly commendable." With that testimonial Turner gained admission to the denomination and began a long relationship with Bishop Payne that would take many twists and turns.

For the next five years Turner led churches in Baltimore and Washington, D.C. But in 1863, with the Civil War cannons roaring and Union casualties running high, Turner's ministry became part of the Union war effort. President Lincoln had already signed the Emancipation Proclamation, and with the desperate need for more troops, sentiment began to shift in favor of putting African Americans in the blue Union uniform. Turner jumped at the chance to strike a blow against slavery and began to recruit black soldiers in the yard of his church. Once the unit was mustered, he volunteered to be commissioned as its chaplain.

The Civil War was not the first American conflict in which blacks were utilized as combatants, but never before had African Americans been commissioned chaplains. These men played a crucial role in the spiritual and material maintenance of the black regiments. Each regiment had about 1,000 men, one of whom was a chaplain. There were about 180 chaplains servicing black regiments during the Civil War, and of them only 14 were black. The chaplain's duties exceeded the role of pastor to the troops. Often he was responsible for ensuring the wounded received medical attention, holding prayer meetings, making sure that soldiers' pay was forwarded to family members back home, visiting the sick, and burying the dead. As the only black commissioned officers in their regiments, these chaplains acted as go-betweens for the black troops and their white commanding officers.

For Turner the Civil War was not only about preserving the Union and ending slavery. It was also about ending spiritual corruption in the souls of white segregationists. Even though he was uncomfortable with war, he preached that this war was righteous. "Disloyalty, traitorism, tyranny, and opposition are all the fruits of one corrupt tree," he wrote in the AME newspaper the *Christian Recorder*, "and the God of heaven has commissioned his holy imperial watchers to hew it down." He was not so naïve as to think that a Union victory would put blacks on equal footing with whites, but he did believe that victory for the North would be a sign of God's will for America and black people. His daily message to his troops was that God wanted the United States to live up to its calling as a nation set apart to be a beacon of racial equality, liberty, justice, and Christian civilization for the world. On the first anniversary of the Emancipation Proclamation, Turner used the sermon to give his assessment of why God had allowed Africans to be stolen from their homeland and enslaved:

Therefore, God, seeing the African stand in need of civilization, sanctioned for a while the slave trade—not that it was in harmony with his fundamental laws for one man to rule another, nor did God ever contemplate that the negro was to be reduced to the status of a vassal, but as a subject for moral and intellectual culture. So God winked, or lidded his eyeballs, at the institution of slavery as a test of the white man's obedience, and elevation of the negro. The extremities of two colors, white and black, were now to meet and embrace each other and work out a great problem by the sanction of Heaven for the good of mankind.

Henry McNeal Turner believed God wanted African Americans to be free; free to serve the God who had called them out of Africa, and free to live as humans created in the image of God.

Not all blacks agreed with Turner's belief that slavery and the war were the work of God's hand. Some suspected that they were the work of the Devil. Turner found himself having to answer the skeptics who called God's character into question. They wanted to know what kind of God allowed cruelty, rape, and the death of so many, occasionally even having a brother from the South kill his brother from the North. As their spiritual leader, Bishop Turner invited his listeners' angry questions. The only comfort he offered was his belief that God sometimes placed stumbling blocks along life's road, both to test and to strengthen the faith of believers. In the end, he argued, such suffering was redemptive. For just as Christ had to suffer for the sins of those yet unborn, America had to suffer for the sake of generations to come. The South needed to be saved from itself, Turner preached, and the North had to accept responsibility for having allowed this sin to take root.

Turner's view of the war as a fight over white moral corruption did not end when the South surrendered. His regiments had performed heroically. And Turner had buried his men with stirring words that spoke of their dedication to God's vision of black people raised up in equality by a new America. As the nation turned its attention toward healing itself and rebuilding its fractured body politic, Turner was pained to see that the evil of racism had outlasted the war. Now in the aftermath of the war, the question stood large of how the church would handle the issue of racial equality. Was the white Methodist Church willing to integrate millions of black people into a new body that would end the privileged treatment of white skin?

As Turner was tormented by questions about the possibility of a unified Methodist Church, his native South saw carpetbaggers arrive from the North ready to capitalize on whatever political and economic opportunity they could find. Northern Christians also came South. Most were filled with good intentions and plans to help rebuild Southern religious life by providing funds to erect new churches and start new schools.

All the major Northern religious groups established missionary efforts in the South after the war. The AME Church had previously been a largely Northern organization, since white Southerners banned it from most of the South for fear it might offer a base to slaves trying to stir rebellion. But after the war the AME Church, under the leadership of Bishop Payne, rushed through the South to build more black Methodist churches. Payne had continued as Turner's mentor, and the younger man still craved the elder bishop's approval the way a son seeks a father's embrace. For his part, Bishop Payne pinned his hopes for the future of the denomination on Turner's energy and determination. But soon after the war's end, a rift between the two men surfaced that would only widen with time.

Payne had been born in Charleston to free black parents who had enough money to educate him. In 1828, at the age of seventeen, he established his own school but had to close it because of state laws that made educating blacks a crime. He tried to operate the school privately,

LOTT CAREY:
A PIONEER MISSIONARY TO AFRICA

The Lott Carey Baptist Foreign Mission Convention, America's first independent black foreign mission society, tackles educational, health, social, and spiritual concerns facing individuals in parts of Africa, India, and the West Indies. Although the mission is deeply committed to planting the seeds of Christian dogma, it is also fiercely committed to helping poorer countries limit the spread of AIDS, improve literacy, and feed the hungry. The Lott Carey Convention now boasts a membership of more than 3,000 churches and approximately 100 full-time workers overseas.

The vision for an all-black mission group grew from the courageous and compassionate heart of a former slave. The self-taught Lott Carey wanted to liberate Africans in his homeland across the Atlantic Ocean. In his mind, Christianity possessed the tools through which the chains of physical and spiritual oppression could be dismantled. In addition, he believed African Americans were positioned, given their history of battling racism in the United States, to help liberate the peoples of Africa. Carey devoted his adult life to improving the social and spiritual conditions of Africans. His commitment to Africa was not truly altruistic. In Africa more than in the United States, he believed, the people respected his efforts and saw him as a human being.

From the beginning, Carey showed signs of his far-reaching vision. Born in Charles City County, Virginia, in 1780 to devout

Baptists, he entered the world a slave, the property of William A. Christian. The road that would lead him to Africa started at First Baptist Church in Richmond, where he was baptized in 1807. From baptism he felt he arose without the markings of a slave. He walked away from the experience looking toward acquiring *complete* liberation—spiritual and physical. Soon called to the ministry, Carey honed his ministerial gifts at the First African Baptist Church. There his vision seemed to widen and deepen. Indeed, his growing yearning to spread Christianity's "liberative" message led him to ask more and more questions about the spiritual and social conditions of people in Africa and in 1815 to organize the Richmond African Missionary Society. His effort was seen as a critical development in the eventual rise of mission work in Africa. During the group's emergence, the United States had been bandying about the idea of a back-to-Africa movement. In some ways Carey's work fostered discussions on blacks' immigration to Africa.

Having raised nearly $1,000 for his first mission trip to Africa, Carey, along with his family and more than twenty other individuals, left Norfolk, Virginia, on the *Nautilus* in March 1821. Forty-eight days later, Carey landed in Freetown, Sierra Leone, where he spent the remaining seven to eight years of his life pursuing his mission work.

What is now named the Lott Carey Baptist Foreign Mission Convention has its roots in a meeting in the mid-1860s in Abyssinian Baptist Church in Harlem, New York. Clergymen from the New York region met to discuss the role of black mission work in general and the type of relationship blacks should hold with the white-run American Baptist Missionary Convention. It was not until black Baptists met at Shiloh Baptist Church in Washington, D.C., in 1897—a few months after the National Baptist Convention, U.S.A., had met in Boston—that the pieces needed for developing a unified black mission group fell into place. The meeting, with twenty-eight preachers, led to the Lott Carey Convention, which eventually became the Lott Carey Baptist Foreign Mission Convention. The move, many believed, united black Baptists in the United States and provided them with the autonomy needed to carve out a strategy for mission work that mirrored their specific theological and political concerns. The goals of the newly formed convention included fostering the "spirit" of foreign mission among local churches and encouraging pastors to increase their involvement with mission work.

In time the convention stretched its reach throughout Africa, in countries including Zaire, Kenya, and Nigeria. It also established itself in Russia, Haiti, Jamaica, and India. Today the convention's biggest initiatives are in health-related concerns, especially confronting the spread of HIV, the virus that can lead to AIDS.

but he was run out of town by whites in 1835. He moved North, and in 1840 he opened a school for black children in Philadelphia, and he became associated with the AME Church a year later. He made the most of his time away from the South by studying at Evangelical Lutheran Seminary in Gettysburg, Pennsylvania, and rapidly climbing through the AME ranks. Payne gained respect among AME clergymen because of his strong education and in 1852 was elected a bishop. His status in the church and friendship provided a stepping-stone for the younger, sometimes abrasive Turner.

In 1865, nearly thirty years to the day after he had left the South, Payne sailed into Charleston harbor accompanied by a few missionaries and established the South Carolina Conference of the AME Church. It was a triumphant occasion that signaled the replanting of AME roots in Southern soil. Like Ezra of the Old Testament, Payne was to form a deep religious culture among people who had been subjugated for generations. At the same time Turner was appointed presiding elder of the Georgia Conference.

The two men appeared to be on the road to achieving all that God had in store for them, yet almost immediately they disagreed over the nature of religious reform in the South. For Payne, the key to the freed slave's spiritual maturation and social elevation was education—most important, an educated leadership. He thought it made little sense to send missionaries throughout the South planting congregations that had to depend on a clergy without formal schooling or theological training. Turner, by contrast, believed Payne's view was unrealistic and ignored the vitality of slave religion in the Deep South. He established congregations headed by those who had been religious leaders on plantations and estates before Emancipation. It was a bonus, he thought, if these church leaders could read and write. As long as they could pray, he argued, as long as they sang spirituals and hymns and, most important, as long as they respected the bishops in charge of the church and truly cared for the people in their pews, there was plenty of time to provide them with literacy.

Under Turner's leadership, the AME Church in Georgia flourished. Because of his skills and experience, and also his charisma and eloquence, he became deeply involved in Georgia politics. In 1866 he

A schoolhouse established to educate and train "colored" children
in Charleston, South Carolina, in 1866. Schools emerging before
the turn of the century were normally aligned with religious organizations such
as the National Baptists or the African Methodists.

was a delegate to the Georgia black convention and worked briefly
as an agent for the federal government's Freedmen's Bureau, set up to
help the freed slaves. Turner also was appointed postmaster of the
Macon post office, a political patronage job often given to blacks during
Reconstruction.

In 1867 he was a Bibb County delegate to the state constitutional
convention, and a year later he was elected to the Georgia House of
Representatives. But just as he had become frustrated with the white

Methodist leadership, Bishop Turner became frustrated with white political leaders in the state and federal governments. Even as an elected official, he felt he was not treated as an equal with white politicians. His anger hit a high point when white state leaders denied him and the other elected black members of the House the seats they had won. Turner led a protest against the ejection, stating, "I am here to demand my rights, and to hurl thunderbolts at the men who dare to cross the threshold of my manhood." In 1870 Turner and the other black state representatives were reseated by an act of Congress.

Not only did the wheels of political change grind slowly, but Turner found electoral politics required a degree of compromise that often diluted the moral principles he thought were important. Moreover, he learned firsthand how cruel and vindictive politics could be. The whites who lost the war were not about to sit idly by as Congress forced Negro equality upon them. They did all they could to discredit the leadership and character of Turner and the other black representatives. Turner was forced to answer fabricated charges of mismanagement of funds at the post office. The most damaging accusation, however, was that of immorality. Rumors began to fly that Turner was having an affair with a woman some identified as a prostitute. Although he never admitted guilt, he offered only weak denials. The damage to his reputation cost him politically. Meanwhile, his already souring relationship with Payne, known for his high moral standards, continued to deteriorate under the weight of the charges.

By 1877 the Reconstruction agenda of the Radical Republicans was in tatters, federal troops no longer enforced the political guarantees that had been made by Southern whites, and the Jim Crow culture of separate and unequal was a fact of life. Lynching was a tool of white terror in the effort to show blacks their proper place in the New South. All these things led Turner to the conclusion that the United States would never be the home of Negroes and that they should seriously consider leaving for a place where African American dignity would be respected. Turner's experience in electoral politics made him skeptical of the possibility for blacks' full inclusion in American society. As his views on the social situation became increasingly radical, Turner concluded that Africa was the land of promise for black people.

African American Christians showed an interest in sending missionaries to Africa for most of the nineteenth century, but Emancipation at home added a new dimension to their work abroad. Many blacks were still trying to make sense of the experience of black enslavement, and some concluded that perhaps God had allowed it to happen so that enslaved blacks could learn about salvation through Christ, then take that message to blacks throughout the diaspora. Some felt that slavery was a part of God's plan for them to gain access to education, high culture, and the rudiments of civilization in order to bring Africa out of the darkness of ignorance and into the age of enlightenment. Turner had never viewed education as the single solution to the many ills that bedeviled freed slaves. He was interested in what Africa had to offer: a home where black skin would not define and limit him. He saw Africa as a possible example of a heavenly kingdom for black Americans, a refuge from racism that would serve as an example for the rest of the "civilized world" of Christian service and goodwill.

In many ways Turner was calling for a second Exodus. The first Exodus for African Americans came via Emancipation, and Bishop Turner believed the second would come via migration to places where blacks could live in dignity, free from the daily terrorism of Jim Crow. Turner was not alone in his views on the limited possibilities afforded blacks in the South, although most chose to remain in the United States instead of migrating to Africa. Between 1860 and 1880 the black population of Kansas swelled from 627 to over 43,000 as blacks sought

The lynching of Charles Mitchell in 1897. Lynchings generally occurred
in front of dozens of people, and often families would share picnic lunches
or dinners while a man or a woman dangled from a rope tied to a tree.

freedom in states and territories that did not share the South's burden
of the past. As they moved westward they established a string of all-
black towns that guaranteed the rights of citizenship to all Negroes.
Places like Nicodemus, Kansas, and Allensworth, California, became
proud testaments to blacks' ability to govern themselves. But Henry
McNeal Turner did not want to go west. He turned his eyes east, across

the Atlantic Ocean, and did all he could to promote mass emigration to Africa. He believed it was the duty of African Americans to Christianize Africa.

Moreover, he thought the continent held infinite possibilities for blacks to build and control their own social, political, and economic institutions. In 1876 Turner was appointed manager of the AME Book Concern in Philadelphia. The position gave him a platform from which to spread his views on religion, politics, and the future of blacks in America. In 1880 he was elected a bishop in the AME Church, becoming one of the first Southerners to hold that title. His appointment was contested by several Northern bishops, partly because

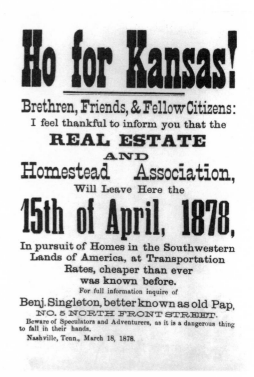

An advertisement encouraging Southern African Americans to settle in the West.

of regional biases and also because of concern about the direction Turner would take African Methodism. True to form, as a bishop he did not shy away from controversy. He immediately advocated the wearing of elaborate clerical vestments and a more demonstrative style of worship, allowing singing in tongues, stomping of feet, and extensive innovation with the hymnal.

Bishop Turner was also an early champion of women preachers. By 1885 he would set the AME Church on its ear by ordaining Sarah Duncan Hughes its first female minister. His commitment to women's empowerment in the church and African missions melded perfectly

African Americans waiting for a steamboat to take them west.

in 1892 with his establishment of the Woman's Home and Foreign Mission Society, an AME auxiliary organization designed to build upon the long-standing tradition of female support for missionary activity. Turner's group was composed primarily of Southern black women, who were effectively excluded from the Northern group's power structure. While he took many bold steps regarding opportunities for women to preach, he never persuaded other black leaders in the church to elevate women to positions of influence in the denomination.

Throughout his tenure as bishop, Turner chose to link the planting of Christianity in Africa with the mass emigration of black Americans to the continent. His insistence on combining the two made him a polarizing figure among both AME laity and clergy. No matter how strongly they felt about the importance of mission work in Africa and no matter how quickly race relations deteriorated in the United States, most African Americans were unwilling to argue for the wholesale abandonment of their American homeland. Turner's perseverance on the issue was rooted in the belief that African Americans were best suited to spread the teachings of Christ to their African brethren. Whites, he believed, were unfit for the task, because even the best of them were fettered by the chains of racism. Only African Americans, he believed, possessed the necessary spirit and education to raise Africans from the burden of superstitions that blocked their path to Christ. In 1895 the bishop's disregard for indigenous African culture even led him to suggest that it was in the best interest of Africans to submit themselves to restricted periods of slavery if they wanted to gain access to the elements of culture and civilization that had lifted the American Negro out of ignorance.

Against a backdrop of controversy, several challenges arose against his leadership in the AME Church. It began when Bishop Turner accepted an invitation to address the National Baptist Convention in Atlanta. Black Baptists had spent the previous three years debating the merits of forming their own national group, so Turner's tough talk against white supremacy undoubtedly appealed to the separatists at the convention. Likewise, he held the Baptists in high regard for their work in Liberia and other parts of Africa, and he saw them as fellow advocates of the idea that the mother continent was key to the future of blacks in America.

It was in this moment of turmoil in his spirit—a desire to go to Africa, where he felt he could finally be free, and a contrary desire to build a black world in America—that Bishop Turner stood in front of the convention in 1895. He declared that "God is a Negro" for a black man in Africa or in America. And Turner's words proved a perfect fit for the moment. The idea he expressed lit a flame in black religion in America at a time when segregation cast a deathly darkness over blacks' hopes of equality. At a time when they needed to hear that they were God's own children, created to do God's will and to bring the light of Christianity to dark places, Turner was there to assure them that they were on the path to righteousness. His words helped galvanize black Baptists' sense of self-determination and invigorated their efforts to distinguish themselves as a unique people in the large Baptist fold.

While some may have found Turner's insistence on God's blackness excessive, they understood the sentiment behind the message: God is on the side of the downtrodden and, despite the bleakness of the American racial landscape, black faithfulness in the face of adversity would be rewarded. With Turner's words in their minds and his commitment to racial solidarity on their hearts, the members of the National Baptist Convention took their first steps toward the uncertainty of the twentieth century, knowing only that God would be there to face with them whatever challenges came their way.

The Business of Religion

Patience is a great gift from God. But as Elias Camp Morris waited in Helena, Arkansas, to receive a package from Richard H. Boyd, corresponding secretary of the National Baptist Home Mission Board, he could only pray to God for more patience. The board had instructed Morris and Boyd to coauthor Sunday school literature for black Baptists nationwide. The men had to have the booklets done by January 1, 1897, and Morris feared that delays on Boyd's end could cause them to miss the deadline and kill the project.

Initially Boyd had doubted they could achieve their goal in the allotted three months. Even if it were possible, he told Morris, it was unlikely that they could garner sufficient support from local congregations to make black Baptist publishing a success. As the black Baptist Church had grown after the Civil War, the churches of former slaves had used texts printed for white Baptists, often with pictures of whites

George Washington Carver with his students at Tuskegee Institute.
Dr. Carver, an agricultural chemist born of slave parents in 1864,
produced over one hundred products from sweet potatoes and
peanuts, such as peanut butter, shampoo, and instant coffee.

illustrating God's word. Although Boyd, who lived in Nashville, fully supported the idea of church texts published for black churches, he told Morris that they had to delay producing even Sunday school literature until it was clear they had enough churches willing to buy. After all, black Baptist Sunday school literature was a direct challenge to the American Baptist Publication Society, a white group in the North that dominated church publications.

Morris and Boyd were operating on a shoestring budget; the home board had given them little money. The wealth they could claim was their wealth of spirit. It was the promise that black Baptists from Massachusetts to Mississippi had faith that they could complete the task. "Our duty is plain," Morris told Boyd, "it is to get out a Sunday school matter by January 1, 1897. This duty has been imposed by the National Baptist Convention, and cannot be set aside by us." Morris was thrilled in late December 1896 as he peeled open the parcel from Boyd containing the National Baptist Convention's first run of Sunday school literature. Even though he knew black church publishing was in its infancy and would have to crawl before it could stand on its own, this batch of Sunday school literature represented a victory of biblical proportions—a black Baptist denomination becoming independent of the white Baptist Church. It was a coming-of-age for black people, who were once thought to possess neither the soul nor the intellect necessary to sustain their own church.

The National Baptist Convention was typical of black religious organizations at the dawn of the twentieth century: it was weak. The question facing every black attempt to organize an independent church was simple and daunting: How do a people just one generation away from chattel slavery find the political, educational, and economic resources to create strong institutions? Even with a deep faith in God, it seemed impossible for the former slaves to build any institutions without the support of whites. Northern Christians—mostly white but

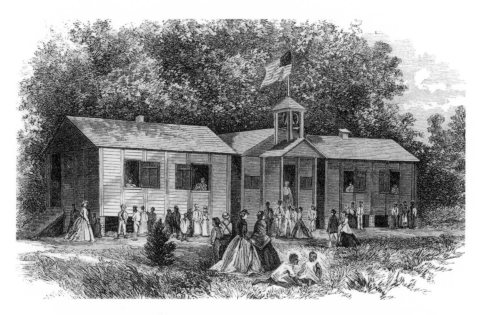

The Penn School established by the Quakers on St. Helena Island, South Carolina.

with some blacks among them—had rushed to the South after the Civil War as missionaries to improve the religious and educational welfare of freed slaves. This movement was labeled the Christian Reconstruction by scholars because it occurred as the federal government was trying to reconstruct the postwar South. Armed with the Holy Bible in one hand and schoolbooks in the other, these Northern missionaries cast their nets to the South, fishing for the souls of the newly emancipated black population. Coming with money and experience, the Northerners hastily built churches to meet the immediate needs of Southern blacks. But there were vast spiritual, educational, social, and political needs in black Southern communities. The struggling churches were completely dependent on the patronage of Northern denominations.

Some of the religious groups that came South after the war got better results than others. For example, the American Missionary Association, which was affiliated with the Congregational Church, succeeded by focusing on providing quality schools for the former slaves. It establish so many schools for blacks in Georgia that Bishop Henry McNeal Turner once doubted aloud whether schooling for blacks in his state would be worth much without the support of the AMA. However, their success did little to promote the development of black Congregational churches in the South. The problem was the Congregationalists lacked tolerance for black folk religious traditions.

As a result, on Sundays the freed slaves stayed away from Congregrationalist services. Instead they went to the Baptist Church, where the singing and testifying fit well with black American traditions. The American Baptist Home Mission Society started working with former slaves in the South in 1862. Southern Baptists had already done a great deal of work among the slaves. When the Southern Baptist Convention was organized in 1845, black Baptists already outnumbered white Baptists in the South. And the number of blacks marching into the Baptist Church grew with the success of the Baptist Home Mission Society's outreach to the slaves, who joined a church that embraced their emotional worship. Not only did Southern Baptists have experience in evangelizing African Americans but they also had a history of interracial worship. Still, worship together did not necessarily imply equal status. One example of the inequality of black and white Baptists was evident in Alabama. At the end of the Civil War, the First Baptist Church of Montgomery had nine hundred members, six hundred of whom were black. Nonetheless, black members did not have full rights and privileges. They sat in the back pews and held no positions of leadership.

The Southern Methodists also struggled with allowing blacks to assume leadership. But black and white Methodists found an answer to the problem in 1870 in Jackson, Tennessee, by creating an independent Colored Methodist Episcopal Church, made up of former black members of the Southern branch of the Methodist Episcopal Church.

Many black Methodists left the churches of their masters after the war and joined Methodist missionary churches from the North. Within a year after the Civil War, only about 78,000 of 208,000 black members in the Southern branch of the Methodist Episcopal Church remained with the original church.

Joining the black church was a difficult decision for many former slaves. The Southern Methodist Episcopal Church was the church of their birth. There was also the question of whether the new black church had a sufficient political and financial base to remain viable and grow. Some Northern abolitionists provided strong financial support to the struggling black churches. But joining an abolitionist denomination provoked anger among more than a few Southern whites. For many blacks it made more sense to work with white Southerners, with whom they had a long, albeit unbalanced, relationship, rather than risk creating racial animosity with their white neighbors.

Nonetheless, black Southern Methodists craved independence. In the tortured politics of the post–Civil War South, black-white relations were understandably in flux. Blacks had expectations of new independence and power. Whites were holding tight to traditions that preserved their privilege. Some whites among the Southern Methodists helped blacks establish segregated congregations within the Methodist Church in order to calm rising tensions over the absence of black leadership in the church. Some blacks had formed the independent African Methodist Episcopal Church, which had growing clout in the South. If that denomination brought its fusion of religion and politics to the region, it would become increasingly difficult for whites to maintain social control over emancipated blacks.

The Reverend Lucius H. Holsey, a Colored Methodist Episcopal bishop from 1873 until his death in 1920, believed that blacks needed to appease their white brethren in order to be free from white anger, which could lead to violence, and to ensure some financial and political support. "Friction in church or state cannot be productive of good to [the Negro] and his children," he said, "and we think it is a legitimate part of Christianity to ameliorate and soften those cruder conditions under which he finds himself as an element in society; hence, we seek the friendship of all, and especially and particularly the fatherly directorship of the Methodist Episcopal Church, South."

Holsey and the Colored Methodist Episcopal Church held their first general conference on December 16, 1870. They established a policy of prohibiting political activity on church grounds, again to quiet white fears of a politically threatening black church. As a result, the Southern Methodist Episcopal Church was sufficiently at ease with the independence of their black brethren that they transferred ownership of some of their property to congregations in the CME Church. The first seven bishops in the church were former slaves, and in its initial century of existence more than 90 percent of its bishops were born in the South, making the church distinctly Southern in its roots and notable for its friendly relations with Southern whites, especially with the Southern Methodist Episcopal Church. The Colored Methodists' social and political conservatism lasted well into the twentieth century. But the decision to exchange Christian political activism for church property remained a point of controversy even as the church grew. It became a major point of contention during the Civil Rights movement of the 1960s.

The relationship between former slaves and former masters was also key to the development of the largest black church in the South, the Baptist Church. For example, Elias Camp Morris, the first president of the National Baptist Convention, took great pride in the mutually supportive relationship his church enjoyed with white Southerners. In reflecting on the experience of newly freed slaves, he wrote: "The

SIGNIFICANT ELECTION SCENE AT WASHINGTON, June 3, 1867,—[Sketched by A. W. M'Callum.]

An African American casting his vote in the Washington election of 1867.

historian is yet to write the true story of the deep and abiding affection which existed between many of the slave owners and their slaves."

Morris's personal experience with slavery was key to his successful leadership of the black Baptist convention. Born into slavery on May 7, 1855, in Springplace, Georgia, he had an unusual, even positive, relationship with his owners. His father, James Morris, ran the master's blacksmith shop. He stayed in town for days at a time and was in charge of handling the money earned at the shop. This meant that James Morris was envied by most slaves and even some whites. When he came home he was able to teach his children to read, to write, and to do basic math long before Emancipation. James Morris also displayed for his children a respectful, even loving relationship with the white owner of the plantation, John Morris. Elias Camp Morris later gave a sense of the extraordinary relationship between his father and his owner when he described his family's departure from the plantation at the end of the Civil War.

> They [Elias's father and master] were both strong men, and yet they wept like babes. My father's parting words were "John I have got to leave you, but you have enough to live on the balance of your life, but I am now getting old and have a large family and have to commence life anew." And like Johnathan [sic] and David they wept (for they were half brothers) until out of sight of each other.

After leaving the white Morris estate, the black branch of the family operated a small farm on the edge of Stevenson, Alabama, until James and his wife, Cora, died in 1869 and 1871, respectively. Shortly after his father's death young Elias was apprenticed to Robert Caver, a shoemaker and Baptist minister. Morris learned both trades and was licensed to preach shortly thereafter. With both of his parents dead, there was little keeping Morris in Alabama. So in 1877 he went west to Helena, Arkansas. Helena was probably not his first choice for relocation; in fact, there is some indication that he had "Kansas fever." Stevenson was about ten miles from the Alabama-Tennessee border, and Morris was familiar with the exodus to Kansas that gripped the African American imagination throughout middle Tennessee.

But although Kansas was heralded as a promised land for blacks, Helena possessed sufficient opportunities to suit Morris's ambitions. By the time he arrived in the Delta Region on the west bank of the Mississippi River, Helena's African American population was growing rapidly. Many blacks had remained in the small town after the war, drawn to the relatively high wages. Planters there were paying double the wages in South Carolina or Georgia. Several counties surrounding Helena were predominantly black. The combination of black soil and black people provided opportunities for professional and economic growth, and Elias Morris was poised to meet the challenge.

Within a year of his arrival, Morris was active in religious enterprises in Arkansas. He connected with the First Baptist Church of Helena, joined the East Arkansas Baptist Association in 1878, and that same year was elected its secretary. Just one year later he was called to the pastorate of Centennial Baptist Church, succeeding his friend and mentor, the Reverend Robert Caver. His organizational skills quickly became apparent, and in the fall of 1879 he organized the Phillips, Lee & Monroe Counties Baptist Association.

In 1880 Morris affiliated with the Arkansas Baptist State Convention, one of two black Baptist factions in the state. One faction was led by two freeborn Northerners; the other was under the leadership of former slaves and Southerners. Throughout the South there was a slow but steady shift in black communities toward leadership indigenous to the region. Morris cast his lot with the former slaves and quickly rose through the ranks. First he succeeded Wesley F. Graham as secretary of the Arkansas Baptist State Convention; two years later, in 1882, he was elected president. Within six hours of his election, Morris negotiated the reorganization of the two factions of black Baptists into one Arkansas Baptist State Convention. The lure he offered was that they could make money.

The first order of business for the new body was to establish a weekly newspaper. Morris viewed publishing as key to the new group's survival. It would provide not only a voice for the leadership but, more important, employment opportunities and revenue for the convention. Under Morris's leadership the *Arkansas Times* (later called the *Baptist Vanguard*) published its first issue in September 1882.

Morris had argued that the lack of a national black Baptist publication gave ammunition to critics who said the heavily Southern denomination of former slaves was made up of stupid, illiterate people. After all, the black Methodists had been operating the AME Book Concern since 1818, and the AME Zion Church had been publishing Sunday school literature for its parishioners since 1841. As black Baptists moved north, their lack of publications became a glaring point of concern. Richard DeBaptiste, pastor of Olivet Baptist Church in Chicago, expressed the view of most black Baptists on the issue when he stated, "Shall we, the foremost in numbers, and the peers of any in talents and culture, continue to be conspicuously wanting in this arena? Can we, as a denomination, a million and a half strong, with forty-one educational institutions, and an army of able educators, advanced scholars, and able writers, afford not to have a Baptist Magazine? We think not."

Morris also saw education as key to the moral and financial future of black Baptists. He quickly established Arkansas Baptist College. Benedict College in South Carolina, Wayland Seminary in Virginia,

Elias Camp Morris and his family. Morris founded Arkansas Baptist College in 1884 and was elected president of the National Baptist Convention in 1895. He was active in the Baptist movement throughout his life.

Selma University in Alabama, and Augusta Institute in Georgia (now Morehouse College) are all examples of black Baptist initiative in the post-Reconstruction South. The primary purpose of these institutions was to provide black Baptists with a literate and well-informed clergy. Many of the teachers Morris hired were not well educated and could boast only of being gifted preachers, but they were the best he could find.

The progression of Elias Camp Morris from former slave to local and regional religious leader coincided with an important shift in the cultural rules of the South. After the Civil War, Northern elites, both black and white, held key political, social, and religious posts in black communities. However, over time black Southern Christians began to generate their own leaders and resent the patronizing attitude of Northerners. Morris was at the cutting edge of this homegrown class of black Southern leadership. And he filled leadership positions below him with fellow slave-born Southerners.

The same sorts of issues that affected black Baptists on local and regional levels were of great concern on the national level. Self-help and cooperation with whites occupied much of the black Baptist agenda. Moreover, by the mid-1880s Jim Crow culture was firmly planted under the guise of a "New South"—a South that felt the best way for whites to coexist with blacks was radically to limit interaction between the groups. At the same time "friends" of the Negro in the North were growing weary of the seemingly small return on their rather large investment in Negro education, and some were beginning to feel that race was a problem the South should handle on its own. With Northerners withdrawing, self-determination became more important than ever to the black church community. The goal for black Baptists was to develop their own national organization, which could draw support from North and South for an institution run by Southerners. Racism had stripped most of the former slaves of their newly won right to vote, so it seemed the key to success was to create a separate political universe in the black church.

At the close of the Civil War, black Baptist membership was an estimated 250,000. By 1890 the number had increased to roughly 1,350,000. Despite this tremendous growth, black Baptists were widely criticized by blacks and whites from other denominations. One minister, William H. Councill, made a career of defending the black Baptist intellect. In one famous letter he contended "whether it be accepted or not, the fact is that there are Negro Baptist ministers in the North and South who, in intellectual ability, in moral power and purity, and in spiritual insight and breadth of wisdom, are the equals of some of

BLACK RELIGION
AND RECONSTRUCTION

The Reconstruction period emerges in American history as one of the grandest yet ultimately disappointing moments for African Americans. From 1865 to 1877 African Americans fought their way through the treacherous valleys of American racism and segregation to some of the highest ranks in the country's local and national political systems. In this period African Americans, yearning to exercise their newly acquired rights as freed women and men, attempted to solidify their positions as citizens as well as to validate their humanity through participation in politics. The personal and political narratives of African American policy makers shed light on the deep impressions they made on the country's political system and on the day-to-day experiences of African Americans throughout the South. Conversely, the history of Reconstruction reveals the extent to which many white Americans tried to impede African Americans. Indeed, through threats of intimidation and acts of violence that led to the deaths of numerous African Americans seeking or holding elected positions, the overall political system supported (through its silence) the demise of the Reconstruction period.

Nevertheless, the period unfolds the rich and complicated history of African Americans in general and the significant role of African American religious leaders' participation in politics in particular. Out of the nearly 1,510 African American office-holders during Reconstruction, some 243, roughly 15 percent, were also ordained ministers from several denominations, including African Methodist Episcopal, AME Zion, Baptist, Methodist, and Presbyterian. African American preachers, moreover, were among some of the most prominent political leaders of the period.

Out of the 243 African American ministers holding political office, the overwhelming majority had been born slaves and, through a variety of means, fought for their freedom. African American ministers held both local and national positions ranging from sheriffs to members of Congress. They occupied 141 seats in state legislatures in the South. Forty-eight of these seats were held by African Americans in South Carolina, 18 in Georgia, 11 in Virginia, and 10 in Alabama. Hiram R. Revels of Mississippi held a U.S. Senate seat from 1870 to 1871, and Jeremiah Haralson of Alabama sat in the U.S. Congress from 1875 to 1877.

But white opposition to blacks' holding political office grew fierce, and African Americans came to find themselves locked out of the political system.

the leading [men and women] of any denomination or race." At the 1890 Fisk University graduation commencement, Booker T. Washington revealed his bias against "uneducated" Baptist ministers when he said that the vast majority of Negro Baptist clergymen lacked education and did not have the proper character to be leaders.

Washington's critique took on extra political power as the black Baptist church became home to a black separatist movement. In 1889 the American National Baptist Convention passed a resolution that urged African Americans to "ask the President of the United States to recommend to the United States Congress an appropriation of $100,000,000 to aid colored people in leaving the South" for destinations throughout the West, where they could form all-black communities.

Ideologically, Morris was part of the vanguard of this separatist movement. He viewed separatism as a way to increase jobs for blacks and asked, "If the large army of Baptist young people who are increasing at such a rapid rate are not encouraged to bookmaking, clerkships, superintendencies, and management by us, whither shall they go as they come forth from the colleges and universities, with religion and learning suited for this very demand?" He added that the effect might be twofold: either the Baptist youth would educate themselves to receive only menial jobs, or they would elect menial jobs over school and then "go backward toward slavery and eternal obscurity." Neither was the desired effect. Hence, employment opportunities that matched the current educational opportunities were crucial. Of course, white racism limited the opportunities available to blacks, so it was hoped that the creation of better-paying jobs for the race would result in lasting self-improvement for black Baptist youths.

This idea led Morris to propose a publishing house that provided blacks an opportunity to demonstrate both to whites and to themselves that blacks were worthy of freedom. "If we do not proceed to do some work which history will rejoice to record, our unworthiness of the good deeds done us will stand out as the most prominent feature in our racial life," he exclaimed. Moreover, "the solution of the so-called race problem will depend in a large measure upon what we prove able to do for ourselves." Morris's thinking about race relations reflected the social conservatism of his denomination. In his mind, blacks had to prove to whites that they were worthy of respect. And a publishing house had the potential to show the best of African American religion and culture. It would confirm to the Northern allies of the black Baptists the importance of having invited in Southern blacks. It was also a chance for Morris to satisfy a deep grievance by showing evidence of black native intelligence.

A longtime advocate of the "business of religion," Morris thought it was essential for the denomination to take on business ventures that would provide not only opportunities for young people but also knowledge of how to use money. In the late nineteenth century, African Americans saw business as a crucial aspect of their maturation process, and Morris and his Baptist comrades were no exception. The publishing house was an opportunity for them to gain respect from whites, although some were concerned that it would be a thumb in the eye of the white Baptists who had aided them for over twenty-five years. To this objection Morris replied that it was humiliating enough that black Baptist books and pamphlets went unpublished because whites did not desire to publish them. He felt they should not be concerned about whether whites would find their work insulting. Such an "earnest and righteous effort" would not fail.

Finally, and probably most significant to Morris, the publishing venture would be a bequest to the generations of black Baptists to follow. Just as "every proud Anglo-Saxon" wants to leave something behind for his or her children, every proud black American should want to leave "a legacy to his posterity." Morris declared that the foreparents of his generation were unable to provide a legacy for their children. This was not their fault, he added, but was a result of "the mistakes of the fathers of this country and the prejudice which like a smoking flax their children will not quench." However, he asserted that it was the duty of his generation to do better for their children than their parents had been allowed to do for them. "He who does not live in the future," Morris declared, "might as well not live at all." He hoped that black Baptists would be able to leave something upon which future generations could build so the denomination could move forward toward full freedom. But in a way his argument for a black Baptist publishing house also reflected the coercive pressures of white racism. Morris's concern over the intellect of former slaves never took into account the damage done by slavery and the denial of education to generations of children.

In September 1895, at the Friendship Baptist Church of Atlanta, Georgia, the various branches of the black Baptist community struggled for unification. The Baptist Foreign Mission Convention, the National Baptist Education Convention, and the American National Baptist Convention came together to form the National Baptist Convention, with Elias Camp Morris as its first president. Although the constitution of the new organization did not provide for a separate publishing house, Morris kept the issue on the table. And with the recent launching of the *National Baptist Magazine* by William Bishop Johnson of Washington, D.C., many more were warming up to the idea of a national Baptist publishing concern. They formed a committee to investigate and report on the feasibility of establishing a publishing house. This was certainly not the first time such a committee was formed, but it was the first time it had a sense of urgency.

On January 7, 1897, Negro Baptist Sunday school literature had its first national distribution. After nearly a decade of struggle, the separatists were able to see their dream come true. One might have expected the matter to simmer down, but in fact the opposition intensified and threatened to consume the National Baptist Convention. Many on both sides of the issue were dissatisfied with the quality of the literature. Some claimed that, with the exception of the "Negro backs" (pictures of blacks on the front and back covers), the literature was no different from that used by whites. This was certainly true, for Richard Boyd had borrowed the printing plates of the Southern Baptist Convention's Sunday School Publishing Board.

RICHARD HENRY BOYD:
BLACK BAPTIST ENTREPRENEUR

Richard Henry Boyd (1843–1922) stands tall in the history of black Baptists in the United States. From the tired hands of the former slave emerged several prominent businesses and long-standing institutions, most notably the National Baptist Publishing Board (NBPB). It was Boyd's tireless efforts, political ingenuity, and entrepreneurial instincts that gave birth to what would become an important and revered institution in black religious history. The rise of the National Baptist Publishing Board is part personal narrative of an unabashedly ambitious and astute individual and part historical narrative of blacks' refusal to accept their marking as second-class cit-

Richard Henry Boyd and his family. Boyd, a former slave,
founded the National Baptist Publishing Board and helped
create the first black Baptist Sunday school literature.

izens and their efforts to construct and maintain their own institutions.

Born a slave on March 15, 1843, on the Gray plantation in Noxubee County, Mississippi, the infant was given a slave name, Richard Gray. Around the plantation, though, the overseers called him Dick, short for Richard. At age ten Dick joined the slaves in the seemingly endless acres of cotton fields. But the land produced neither the quality nor the quantity of products needed to secure big profits and, like other plantation owners in the region during the antebellum period, the Gray family looked west of the Mississippi River for land that could yield greater dividends. In their pursuit of larger profits, the Grays split up many families, including Dick's, and the fifteen-year-old was acquired by a member of the Gray clan in Texas. Dick's mother, Indiana, and siblings moved with another part of the Gray family to Grimes County, Texas.

Dick established a reputation as a prodigious field hand, and the patriarch of the Gray family and his sons had a special liking for Dick. A portion of his time off the fields was spent with the Grays. What Dick acquired, if anything, from these interactions is highly speculative, but some believe they taught him a critically important lesson: Whites do not possess a monopoly on knowledge. In fact, the interactions most likely reinforced his beliefs in the inalienable intelligence and humanity of blacks. He walked away assured that he, too, could manage a successful business if he were not bound by slavery.

Shortly after the Grays moved to Texas, the Civil War began. Dick and countless other slaves stood alongside their masters in the Confederate military. The patriarch Gray and two of his three sons died during the war. When the last Gray son was injured, Dick cared for him as they made their way back to the family plantation. Obviously Dick was aware that the Union was winning; in fact, he saw thousands of slaves head toward the Yankee line to join the Union army. But Dick, for whatever reasons, remained at the side of his master. Back in Texas the twenty-one-year-old was finally reunited with his mother and some of his siblings. He worked and managed the Gray plantation until the deterioration of the economy around it forced the Grays to fold their operations.

After the fall of slavery, Dick began envisioning himself as a new creature. All that he had dreamed as a slave could finally manifest itself. He already had the motivation, intelligence, and experience to actualize his dreams; now he needed only blueprints.

When his first wife died within eleven months of their marriage, the grief-stricken Dick turned to the church. He was baptized in 1869 at Hopewell Baptist Church in Navasota, Texas. He changed his name to Richard Henry Boyd after discovering the roots of his father's family and thereafter entered the ministry.

Boyd grew steadily in prominence among black Baptists in Texas. He helped organize the first black Baptist association in the region. As a leader in the Texas Negro Baptist Convention, Boyd traveled in a wagon throughout Texas preaching a message of unity among black Baptists. From the beginning of his involvement in the political structure of the Baptist Church, he began nurturing his vision of making the church a major institution that reached beyond state borders.

Boyd's efforts at unifying black Baptists both in and outside Texas were fueled in 1893, when the Texas convention divided over the future of the black schools and colleges started by the American Baptist Home Mission Society, the Northern white Baptist group. The ABHMS proposed closing several schools and colleges, but Boyd and his supporters, already fuming over the American Baptist Publication Society's decision to censor the writings and distribution of black church literature, fiercely opposed the proposition. The issue for Boyd was control, and he was aggressively attempting to push all of the church's business and religious affairs out of white hands. For far too long, Boyd proclaimed, the ABHMS had blocked the publication of texts by black ministers, and if black Baptists ignored such moves, real autonomy was in jeopardy. In addition, Boyd saw the extent to which black Baptists made up the ABHMS's revenue base and, since their concerns were not being addressed by the society, he advocated for the creation of an independent publishing company, owned and operated by black Baptists. Boyd soon discovered that undertaking efforts toward financial autonomy meant facing opposition from two groups: Northern white Baptists and some black Baptist ministers.

In 1897, after several arduous years of political maneuvering and negotiating, Boyd and a handful of his supporters founded the National Baptist Publishing Board in Nashville, Tennessee. Its financing and initial support came from Boyd's personal accounts. In its first few years, the fledgling company faced biting opposition. Boyd had to enter a sophisticated, costly, and time-consuming campaign to convince black preachers to support and purchase its publications. In addition, the company had to attract and help train black writers, editors, and machinists to operate the printing plant.

While building the NBPB into a viable institution, Boyd extended his reputation outside black Baptist circles. He was especially visible in Booker T. Washington's National Negro Business League and admired Washington and his ideology. In fact, he often referred to Washington as the "champion of the people." Boyd, like Washington, took every step to encourage blacks to participate fully in the development and acquisition of businesses. He also helped form the Citizens Savings Bank and Trust Company in Nashville, serving as its first president from 1904 to 1922. In addition to being one of the *Nashville Globe*'s leading investors, Boyd was elected the first president of the Nashville Globe Publishing Company in 1906.

Morris's explanation for the "Negro backs and white guts" fell on deaf ears. In fact, it seemed only to strengthen his critics' claims that Sunday school publishing should be left to those with the resources to do it best—the whites. The National Baptist Convention, they argued, needed to focus on winning souls for Christ at home and abroad. Moreover, the American Baptist Home Mission Society had long been a generous benefactor of black Baptist schools in Florida, and many did not feel it proper to upset such long-standing relations with Northern Baptists. They felt the publishing house was being forced on them and, as a result, was dividing black Baptists instead of unifying them. A *Florida Evangelist* editorial read, "If some of our brethren must have an Afro-American Publishing House and nothing else will do them, then let them have it, and let us who are satisfied with out such a publishing house do what we can to make the American Baptist Publishing Society—the publishing house of our choice—all that it ought to become. Afro-American Baptists cannot afford to divide upon so small a matter as the starting of a useless and needless publishing house, which has never been established and which is not destined to be." Divide, however, is exactly what they did.

In September 1897 a group of National Baptists met at Shiloh Baptist Church in Washington, D.C., and formed the Lott Carey Foreign Mission Society (later Convention) as an outlet for those who believed the work of the denomination should be to "go into the world and teach all nations—baptizing them in the name of the Father, Son, and Holy Spirit." Lott Carey was the first African American missionary to Liberia. Many of the congregations that affiliated with Lott Carey maintained an alliance with the American Baptist Publication Society over the National Baptist Publishing Board. But this split did not deter Morris and those who felt that the only way black Christians could walk with dignity was if they had their own houses of worship. That approach allowed them to employ their own children, as well as to write and publish their own hymns and interpretations of scripture. For years to come the face of the black religious landscape would reflect Morris's vision. He created a legacy for generations of black Baptists. And although his publishing venture did not always unify black Baptists—there would be arguments over ideology and money—it did instill race pride and a sense of accomplishment in those who supported the cause. The National Baptist Convention and its publishing board are sterling examples of former slaves using the faith that had never been stripped from them as the cornerstone for separate but equal institutions that reflected their lives.

6

"Saved, Baptized and Holy Ghost Filled"

Lord, as of old at Pentecost Thou didst thy pow'r display,
With cleansing, purifying flame, descend on us today.
Lord send the old-time pow'r, the Pentecostal pow'r!
The floodgates of blessing on us throw open wide!
Lord, send the old-time pow'r, the Pentecostal pow'r,
That sinners be converted and thy name glorified!
—"PENTECOSTAL POWER"

Visitors to the grand spectacle could not quite explain what they witnessed. From the moment they stepped off the train, blocks away from the old livery stable in the downtown section of Los Angeles, they could feel an energy that could only be, they thought, the Spirit of God at work. Long before the building at 312 Azusa Street that once housed the city's first black Methodist congregation was in sight, one could hear the screams, laughter, songs, barks, and shouts of sheer ecstasy from the saints who had been mounted by the Holy Spirit.

312 Azusa Street in Los Angeles, an abandoned African Methodist Episcopal Church building, is where William J. Seymour taught the importance of "baptism" by the Holy Spirit and speaking in "tongues."

William J. Seymour (1870–1922) is considered one of the founders
of the American Pentecostal movement. He helped found the
Pacific Apostolic Faith group and published the *Apostolic Faith,*
a newspaper that discussed Pentecostal doctrine.

Any observer immediately would have noticed many things that
appeared out of the ordinary. First, William J. Seymour, the leader of
the group, was a short, disheveled black man with only one eye, an
odd-looking fellow whose appearance matched the revival's carnival
atmosphere. Nonetheless, he had a magnetic spiritual vision that drew

thousands to the site daily. He did not do very much preaching. Most of the time he sat behind a makeshift pulpit with his head completely covered with a wooden packing crate praying, singing, or murmuring to himself in a language no one else could understand. Occasionally he would walk through the crowd in a trancelike state, oblivious to the countless touches of strangers, not to mention the dollars visitors shoved into his pockets. He remained focused, waiting to receive a word from on high, as the crowd worked itself into a frenzy.

William J. Seymour was born in 1870 in Centerville, Louisiana. His parents were former slaves, and their religious practices reflected the syncretistic experience of religion in the slave quarters. The emphasis of Haitian Voodoo was strong in the region, so even as a child Seymour was comfortable with supernatural encounters with the divine. All of this, combined with his black Baptist heritage, gave him an outlook that was open to multiple avenues of religious expression. He just wanted to get closer to God any way he could.

As a young man Seymour moved to Indianapolis and began to seek a more meaningful walk with Jesus. He decided to leave the Baptist Church and unite with an all-black congregation of the Methodist Episcopal Church, North. Around this time Seymour was stricken with smallpox, losing the use of one eye. Night and day he prayed to God to heal his body and restore his sight, and while he never regained use of the eye, his spiritual vision became clearer as he focused more on his relationship with God. Around 1901 Seymour embraced the teachings of the Holiness movement and traveled extensively, preaching about the sanctifying work of grace.

Approaching Azusa Street, one would also mark the interracial character of those who convened at the revival. From its inception in April 1906 to the time it began to wane in 1909, the revival was made up of Christians of all kinds and colors. The core group was a handful of black Christians who had been voted out of Second Baptist Church in Los Angeles for their belief that one could be made holy in this life. But the revival's black leaders did not prevent others from worshiping with them, and before long many whites, Mexicans, and Chinese came to recognize that they might miss their blessing if they allowed race to keep them away from the event. The *Apostolic Faith*, the mission's newspaper, edited by Seymour, remarked, "It is noticeable how free all nations feel. If a Mexican or a German cannot speak English, he gets up and speaks in his own tongue and feels quite at home for the Spirit interprets through the face and people say amen. No instrument that God can use is rejected on account of color or dress or lack of education. This is why God has so built up the work." Indeed, whites made up a large portion of the throng by the revival's end, but blacks and women played important leadership roles throughout the history of the gathering.

The interracial character of the Azusa Street Revival and Mission stood in sharp contrast to life in the rest of the nation. The year 1906 was greeted by racial violence in Savannah, Georgia, when whites disrupted the local Emancipation Day parade. Two months later Illinois state militia units were called up to suppress riotous whites who had terrorized blacks in Springfield for four days. And by summer's end, racial incidents between black soldiers and white civilians in Brownsville, Texas, had resulted in President Theodore Roosevelt's dishonorable discharge of three black companies. In a time when racism had forced hundreds of thousands of black Americans away from the Deep South, and many more to wonder if blacks could ever stake their claim in such a racist society, the Pentecostal movement offered a glimpse into the Kingdom of Heaven on earth. It gave believers a sense of what life could be like if Christians treated all people as equals. "God recognizes no flesh, no color, no names," reported the *Apostolic Faith*.

"Azusa Mission stands for the unity of God's people everywhere. God is uniting His people, baptizing them by one Spirit in one body."

Probably the most striking characteristic of the Azusa Street Revival was glossolalia, speaking in tongues. Protestant Christians assert that salvation comes through the grace of God and to earn it one can do nothing other than have faith that Jesus' death and resurrection are payment enough for the world's sins. Holiness Christians additionally believe in what is often called the second blessing, or the second work of grace. They assert that not only does God's grace save the believer, but it also has the power to purify and set one free from sin. In effect, through the Holy Spirit, God's grace can sanctify one who actively works toward this blessing. Pentecostals are Holiness Christians who believe speaking in tongues, a third work of God's grace, is a sign that the Holy Spirit dwells in an individual. This baptism by the Holy Spirit is seen as a higher level of spiritual attainment.

According to the New Testament, on the day of Pentecost, Jesus' disciples and others gathered from all parts of the world. Suddenly the Holy Spirit descended on their heads in the form of tongues of fire, and they were able to speak in one another's languages with no prior knowledge of them. In the United States, the phenomenon was limited until this outbreak in Los Angeles. Before then, speaking in tongues in the United States was primarily a matter of people speaking in known languages and dialects. For example, believers would suddenly speak fluent Chinese or French without having studied the language. But now they also spoke in unknown tongues, which could be understood only by God and the occasional person who had the gift of interpretation. Moreover, glossolalia became the impetus for world evangelism among many Christians, who contended that God was readying the world for his return by knocking down the barriers to spreading the gospel. Some even believed that missionaries no longer needed to study

Twi or Arabic or Bengali in order to do God's work in the farthest corners of the globe.

Elias Camp Morris, founding president of the Arkansas Baptist State Convention and first president of the National Baptist Convention, was aware of the Holiness Pentecostal movement and probably was concerned about its perceived anti-intellectual nature, in addition to what some considered its excesses in worship. He lived long enough to see the movement grow from a fanatical religion to a fringe movement with a small but significant impact on every major Christian denomination in America and finally to a global phenomenon. Black Baptists already had a reputation for supporting uneducated clergymen, so the spread of these movements was of grave concern to black Baptist leaders. They believed any association with a movement that seemed to build upon the least-flattering aspects of slave religion (ignorance, superstition, and excessive emotionalism) would be a step in the wrong direction.

When Morris formed Arkansas Baptist College in 1884, the goal was to educate the state's black Baptist clergy with the hope that their knowledge would trickle down to their congregants. By the last decade of the nineteenth century, the college had a seminary-trained black president and academic dean who brought a zeal for religious reform. Little did they know that the college would play a role in the establishment of a religious movement that believed in the necessity of religious education.

Moreover, the church historian Dale Irvin notes, "Many Baptist churches in the south, including the Black Baptist congregations, were concerned with strengthening their institutional identity during this period; and this meant increased attention to denominational distinctives. The holiness movement cut against the denominational grain on the other hand, challenging Baptist identity formation, and calling instead for the unity of the churches." Thus, the Holiness movement offered an alternative religious identity that downplayed theological difference in favor of unity under the reign of the Holy Spirit.

In late 1887 a young man glided into Morris's church in Helena, Arkansas, in search of advice. Charles Price Jones believed God was calling him to preach the gospel in Africa, but he was uncertain about the best way to pursue this mission. Jones knew that Morris would be able to point him toward the resources he needed for his journey, and God would do the rest. Morris was impressed by Jones's way with words and his love of God. There was no doubt this young man had a bright future; all he needed was the proper guidance. Instead of helping Jones find the money and contacts to embark on an African sojourn, Morris told him that the

Boys' Dormitory and President's Quarters

Arkansas Baptist College
LITTLE ROCK, ARKANSAS
INDUSTRIAL   LITERARY   RELIGIOUS
For Negro Youth

Carries all the literary branches from the thorough grammar school course to the conservative college. Gives particular attention to industrial training, such as sewing, cooking and laundering for girls; carpentry, cabinetmaking, house building, gardening, printing and dairying for boys. Other branches of industry will be introduced from time to time as means are supplied by friends to that branch of education.

JOS. A. BOOKER, PRESIDENT

Boys' Industrial Building, for Carpentry, Cabinetmaking, Painting, Printing, Shoemaking, Etc.

Arkansas Baptist College in Little Rock, an industrial school for the training of blacks.

key to unlocking God's plan for him lay to the north—in Little Rock. Morris believed that each generation should exceed the labors and rewards of the last, and without an education Jones and those of his generation were doomed to repeat the errors of the past. So he sent the young man to Arkansas Baptist College and set in motion a providential encounter between two pioneers in the Holiness and Pentecostal movements among blacks.

Charles Price Jones (1865–1949) was a Baptist minister in Alabama and Mississippi before founding what became the Church of Christ (Holiness) USA. He also wrote over a thousand gospel songs. The organization's first Holiness convocation was held in 1897.

Jones was born in Georgia on December 9, 1865. His mother was a former slave who instructed him early in life about the importance of leading a God-fearing life. As a teenager Jones, in search of work, found his way to Arkansas. On Cat Island in the Mississippi River in Arkansas, a scene of mass baptisms, he was converted to Christianity at the age of nineteen. This experience began his quest for a deeper knowledge of God.

It seemed as if God was guiding Jones's steps. After enrolling at Arkansas Baptist College in January 1888, he quickly advanced through the state convention's leadership ranks. Charles Lewis Fisher, the school's academic dean, ordained Jones in October of that year, and by the close of the decade Jones was pastor of three churches. In those days most rural churches could not afford a full-time pastor, so it was not uncommon for a minister either to have an additional profession or to pastor more than one church. Jones graduated in 1891 and went on to be elected corresponding secretary of the state convention and a trustee of the college, not to mention editor of the convention's newspaper, the *Baptist Vanguard.* He left Arkansas in 1893 to become the pastor of Tabernacle Baptist Church in Selma, Alabama.

The year Jones left Arkansas, he met a young man who had also gone to Morris for counsel. Charles H. Mason was twenty-six years old when he made his way to Helena from Plumerville, Arkansas, seeking Morris's wisdom. His marriage had fallen apart because his wife did not approve of his call to preach the gospel. Mason felt his heart had been ripped out by the divorce, and he needed reassurance that serving God at the expense of matrimony was the right thing to do. Morris assured him that following God's call is always correct, and that the storm in his personal life would pass once Mason redirected his energy toward Christ. Like Jones before him, Mason followed Morris's advice and on November 1, 1893, enrolled in Arkansas Baptist College.

Charles H. Mason (1866–1949) started the Church of God in Christ in 1897.

Mason's educational experience proved to be short-lived, however. Unlike Jones, he found little about the school useful. In particular, Mason objected to the higher biblical criticism, which instructed students to place scripture in its historical context in order to get an accurate understanding of its meaning. Mason believed the truth of scripture was timeless, and true believers didn't need a college education to understand it. God's word was clear enough for a child to understand. In January 1894, Mason left Arkansas Baptist College for fear that his faith would become tainted by the rational approach to religion espoused by the seminary-trained faculty.

He preferred that old-time religion of the slaves. Theirs was a faith that did not need to comprehend the poetics of the Hebrew language to know that God was true to his word when he delivered Daniel from a den of lions. Theirs was a faith that did not need to compare the Gospel of Matthew with that of Luke to get a more complete picture of the life of Jesus. All they needed to know was that he died for their sins and his blood cleansed them of all unrighteousness, and one day they would leave the troubles of this world for a better place. And in that place no one would ask their birth status as slave or free, whether they owned property or were sharecroppers, or whether they had enough education to write their names. No, all Jesus would say to them would be, "Come unto me, all ye who are burdened and heavy laden and I will give you rest."

Mason was equally concerned about some Baptists' belief that spiritual progress came only with distance from the worship traditions of slave religion. As America prepared to close the book on the nineteenth century, he had the sense that too high a price was being paid for progress, that too much was changing too fast, and that the church was making a big mistake by following secular society down an efficient yet spiritually vacuous route toward the future.

Jones agreed with many of Mason's concerns, but he was willing to use the tools of modern scholarship to deepen his faith and that of his parishioners. In 1895 he accepted the pastorate of Mt. Helm Missionary Baptist Church in Jackson, Mississippi, and almost immediately he ran into problems over his espousal of Holiness teachings. Although the congregation was impressed by his oratorical gifts and overall learnedness, most did not support the unorthodox (for Baptists) teachings of Holiness. Mason was also in Mississippi, near Lexington, and he, too, had become interested in the Holiness movement. For him and Jones the movement breathed new life into the spiritually dead places in Baptist life and worship.

The two men crisscrossed the state, speaking at local Baptist association meetings and revivals about how the second work of grace brought perfection to the life of the believer. They were having a small yet significant impact on the religious life of the National Baptist Convention, and Morris felt something needed to be done to prevent the spread of Holiness teachings among Baptist churches. It was not that he opposed their growing notoriety and personal influence but that he believed their teachings were not supported by sound scriptural interpretation. If this movement were left unchecked, the essence of what it meant to be Baptist would be eroded by outside beliefs.

In 1897 Jones and Mason held a Holiness revival in a cotton gin near Lexington, Mississippi, and this gathering became, in effect, the organizing session for the first Holiness denomination in the South—the Church of God in Christ. Mason claimed that one day in 1895, while he was walking down the streets of Little Rock, the name of the church was revealed to him by God. Later that same year Mason incorporated the church in Memphis, Tennessee, and became a lightning rod for anti-Holiness rhetoric among black Baptists. Meanwhile, Jones continued to hold annual Holiness convocations at Mt. Helm despite the fact that the movement was having an unsettling effect on the people. Finally, in 1898, after Jones proposed to change the name of the church to the Church of Christ, Holiness, Mt. Helm split into two congregations. The schism did not end the controversy, for Jones was sued over the matter, and it wound up in the Mississippi Supreme Court, which ruled that since the land had been donated for the purpose of establishing a Baptist congregation, Jones had no legal ground to change either the church's name or its theology.

In 1899, at the annual session of the Arkansas Baptist State Convention, Morris set the record straight on Baptist views of perfection. "That there is such a doctrine as sanctification taught in the Scriptures must be admitted. And that this doctrine has been wantonly perverted and misunderstood must also be admitted. But I am charitable enough to say that many who have misunderstood and misinterpreted this doctrine have done so from honest convictions which had formed in their hearts on account of incompetent teachers." He went on to explain proper Baptist doctrine on the subject. "Having received the sanctifying influence of the Holy Spirit in our hearts, we set about a cultivation of it with an anxious desire that we may become more like Christ each day. The deformity which sin has brought on us will only be lost in the regeneration of the world." In Morris's view, then, one would be made holy only at death or when Christ returns to judge the world.

Jones and Mason were undeterred by the criticism from their former mentor. They were not angry with Morris and the others in the convention who disagreed with their views; they simply believed them to be wrong. In fact, they felt they had evidence that the convention was wrong. They were witnesses of the possibilities of sanctification. In characterizing his stormy experience with the Baptists, Jones wrote, "Having reached my decision to follow my convictions and my Lord, I was looked upon as a fanatic by some, by others as of weak brain; by yet others as a sharper trying to distinguish myself by being different, by nearly all as a heretic." Yet he and Mason continued to spread the message of sanctification with the Church of God in Christ as their vehicle.

As the movement grew so did the friction between Mason and Jones. Both were strong men, able preachers, and religious visionaries who believed God had special work for them. Once word of the Azusa Street Revival reached the two men, their doctrinal differences began to surface. Ironically, after he had been run out of Mt. Helm and expelled from state and national Baptist organizations for preaching and teaching about Holiness, it was Jones who was most resistant to the ideas of Pentecostalism and glossolalia. While he was initially supportive of speaking in tongues, he did not think it was the only way to receive the Holy Spirit. Years later he wrote, "Now Christ and the Holy Spirit are one, and the Holy Ghost is the Spirit of Christ (I Pet. 1:11), and when one who has asked and received the Holy Ghost (Luke 2:1−3) lets some sort of Spirit cause him to deny the witness of the Spirit that he has received in order to get that other spirit, I ask you what has he done."

Although Mason was unconvinced, he was willing to learn all he could about the doctrine and test its scriptural support. In 1906 curious observers descended on Los Angeles from various regions of the country representing every Christian tradition imaginable. The National Baptist Convention was scheduled to have its annual meeting in Los Angeles in September 1906, but for unstated reasons it changed locations. Perhaps there was concern that convention delegates would be drawn to Azusa Street and its teachings. The convention was just beginning to recover theologically from the rumblings caused by Mason and Jones, and the leadership was determined to prevent an uncontrollable outpouring of the Holy Spirit that would engulf their traditions. Mason, however, was very interested in seeing for himself what was taking place on the West Coast. He and Jones were familiar with William Seymour from the Holiness revivals he'd run in 1905 throughout Mississippi, but at that point the third work of grace, speaking in tongues, had not been as prominent a feature of Seymour's teachings. Mason was unable to persuade Jones to accompany him to Los Angeles, but in March 1907 two other Church of God in Christ ministers, D. J. Young and J. A. Jeter, agreed to join him.

Once in Los Angeles, the three men went directly to Azusa Street to see if they could get a taste of the movement that was sweeping America's religious landscape. While Mason was open to baptism by

THE APOSTOLIC FAITH

"Earnestly contend for the faith which was once delivered unto the saints."—Jude 3.

Vol. 1, No. 1 | Los Angeles, Cal., September, 1906 | Subscription Free

Pentecost Has Come

Los Angeles Being Visited by a Revival of Bible Salvation and Pentecost as Recorded in the Book of Acts

The power of God now has this city agitated as never before. Pentecost has surely come and with it the Bible evidences are following, many being converted and sanctified and filled with the Holy Ghost, speaking in tongues as they did on the day of Pentecost. The scenes that are daily enacted in the building on Azusa street and at Missions and churches in other parts of the city are beyond description, and the real revival is only started, as God has been working with His children mostly, getting them through to Pentecost, and laying the foundation for a mighty wave of salvation among the unconverted.

The meetings are held in an old Methodist church that had been converted in part into a tenement house, leaving a large, unplastered, barn-like room on the ground floor. Here about a dozen congregated each day, holding meetings on Bonnie Brae in the evening. The writer atended a few of these meetings and being so different from anything he had seen and not hearing any speaking in tongues, he branded the teaching as third-blessing heresy and thought that settled it. It is needless to say the writer was compelled to do a great deal of apologizing and humbling himself to get right with God.

In a short time God began to manifest His power and soon the building could not contain the people. Now the meetings continue [illegible] is kindling all over the city and surrounding towns. Proud, well-dressed preachers come in to "investigate." Soon their high looks are replaced with wonder, then conviction comes, and very often you will find them in a short time wallowing on the dirty floor, asking God to forgive them and make them as little children.

It would be impossible to state how many have been converted, sanctified and filled with the Holy Ghost. They have been and are daily going out to all points of the compass to spread this wonderful gospel.

———

BRO. SEYMOUR'S CALL.

Bro. W. J. Seymour has the following to say in regard to his call to this city:

"It was the divine call that brought me from Houston, Texas, to Los Angeles. The Lord put it in the heart of one of the saints in Los Angeles to write to me that she felt the Lord would have me come over here and do a work, and I came, for I felt in was the leading of the Lord. The Lord sent the means, and I came to take charge of a mission on Santa Fe Street, and one night they locked the door against me, and afterwards got Bro. Roberts, the president of the Holiness Association, to come down and settle the doctrine of the Baptism with the Holy Ghost, that it was simply sanctification. He came down and a good many holiness preachers with him, and they stated that sanctification was the baptism with the Holy Ghost. But yet they did not have the evidence of the second chapter of Acts, for when the disciples were all filled with the Holy Ghost, they spoke in tongues as the Spirit gave utterance. After the president heard me speak of what the true baptism of the Holy Ghost was, he said he wanted it too, and told me when I had received it to let him know. So I received it and let him know. The beginning of the Pentecost started in a cottage prayer meeting at 214 Bonnie Brae."

———

LETTER FROM BRO. PARHAM.

Bro. Chas. Parham, who is God's leader in the Apostolic Faith Movement, writes from Tonganoxie, Kansas, that he expects (D. V.) to be in Los Angeles Sept. 15. Hearing that Pentecost had come to Los Angeles, he writes, "I rejoice in God over you all, my children, though I have never seen you; but since you know the Holy Spirit's power, we are baptised by one Spirit into one body. Keep together in unity till I come, then in a grand meeting let all prepare for the outside fields I desire, unless God directs to the contrary, to meet and see all who have the full Gospel when I come."

———

THE OLD-TIME PENTECOST.

This work began about five years ago last January, when a company of people under the leadership of Chas. Parham, who were studying God's word, tarried for Pentecost in Topeka, Kan. After searching through the country everywhere, they had been unable to find any Christians that had the true Pentecostal power. So they laid aside all commentaries and notes and waited on the Lord, studying His word, and what they did not understand they got down before the bench and asked God to have wrought out in their hearts by the Holy Ghost. They had a prayer tower from which prayers were ascending night and day to God. After three months, a sister who had been teaching sanctification for the baptism with the Holy Ghost, one who had a sweet, loving experience and all the carnality taken out of her heart, felt the Lord lead her to have hands laid on her to receive the Pentecost. So when they prayed, the Holy Ghost came in [illegible] an unknown tongue. This made all the Bible school hungry, and three nights afterward, twelve students received the Holy Ghost, and prophesied, and cloven tongues could be seen upon their heads. They then had an experience that measured up with the second chapter of Acts, and could understand the first chapter of Ephesians.

Now after five years something like 13,000 people have received this gospel. It is spreading everywhere, until churches who do not believe backslide and lose the experience they have. Those who are older in this movement are stronger, and greater signs and wonders are following them.

The meetings in Los Angeles started in a cottage meeting, and the Pentecost fell there three nights. The people had nothing to do but wait on the Lord and praise Him, and they commenced speaking in tongues, as they did at Pentecost, and the Spirit sang songs through them.

The meeting was then transferred to Azusa Street, and since then multitudes have been coming. The meetings begin about ten o'clock in the morning and can hardly stop before ten or twelve at night, and sometimes two or three in the morning, because so many are seking, and some are slain under the power of God. People are seeking three times a day at the altar and row after row of seats have to be emptied and filled with seekers. We cannot tell how many people have been saved, and sanctified, and baptised with the Holy Ghost, and healed of all manner of sicknesses. Many are speaking in new tongues, and some are on their way to the foreign fields, with the gift of the language. We are going on to get more of the power of God.

———

Many have laid aside their glasses and had their eye sight perfectly restored. The deaf have had their hearing restored.

———

A man was healed of asthma of twenty years standing. Many have been healed of heart trouble and lung trouble.

———

Many are saying that God has given the message that He is going to shake Los Angeles with an earthquake. First, there will be a revival to give all an opportunity to be saved. The revival is now in progress.

———

The Lord has given the gift of writing in unknown languages, also the gift of playing on instruments.

———

A little girl who walked with crutches and had tuberculosis of the bones, as the doctors declared, was healed and dropped her crutches and began to skip about the yard.

———

All over this city, God has been setting homes on fire and coming down and melting and saving and sanctifying and baptizing with the Holy Ghost.

———

Many churches have been praying for Pentecost, and Pentecost has come. The question is now, will they accept it? God has answered in a way they did not look for. He came in a humble way as of old, born in a manger.

———

The secular papers have been stirred and published reports against the movement, but it has only resulted in drawing hungry souls who understand that the devil would not fight a thing unless God was in it. So they have come and found it was indeed the power of God.

———

Jesus was too large for the synagogues. He preached outside because there was not room for him inside. This Pentecostal movement is too large to be confined in any denomination or sect. It works outside, drawing all together in one bond of love, one church, one body of Christ.

———

A Mohammedan, a Soudanese by birth, a man who is an interpreter and speaks sixteen languages, came into the meetings at Azusa Street and the Lord gave him messages which none but himself could understand. He identified, interpreted and wrote a number of the languages.

———

A brother who had been a spiritualist medium and who was so possessed with demons that he had no rest, and was on the point of committing suicide, was instantly delivered of demon power. He then sought God for the pardon of his sins and sanctification, and is now filled with a different spirit.

———

A little girl about twelve years of age was sanctified in a Sunday afternoon children's meeting, and in the evening meeting she was baptized with the Holy Ghost. When she was filled those standing near remarked, "Who can doubt such a clear case of God's power."

———

In about an hour and a half, a young man was converted, sanctified, and baptized with the Holy Ghost, and spoke with tongues. He was also healed from consumption, so that when he visited the doctor he pronounced his lungs sound. He has received many tongues, also the gift of prophecy, and writing in a number of foreign languages, and has a call to a foreign field.

———

Many are the prophesies spoken in unknown tongues and many the visions that God is giving concerning His soon coming. The heathen must first receive the gospel. One prophecy given in an unknown tongue was interpreted, "The time is short, and I am going to send out a large number in the Spirit of God to preach the full gospel in the power of the Spirit."

———

About 150 people in Los Angeles, more than on the day of Pentecost, have received the gift of the Holy Ghost and the Bible evidence, the gift of tongues, and many have been saved and sanctified, nobody knows how many. People are seeking at the altar three times a day and it is hard to close at night on account of seekers and those who are under the power of God.

———

When Pentecostal lines are struck, Pentecostal giving commences. Hundreds of dollars have been laid down for the sending of missionairies and thousands will be laid down. No collections are taken for rent, no begging for money. No man's silver or gold is coveted. The silver and the gold are His own to carry on His own work. He can also publish His own papers without asking for money or subscription price.

———

In the meetings, it is noticeable that while some in the rear are opposing and arguing, others are at the altar falling down under the power of God and feasting on the good things of God. The two spirits are always manifest, but no opposition can kill, no power in earth or hell can stop God's work, while He has consecrated instruments through which to work.

———

Many have received the gift of singing as well as speaking in the inspiration of the Spirit. The Lord is giving new voices, he translates old songs into new tongues, he gives the music that is being sung by the angels and has a heavenly choir all singing the same heavenly song in harmony. It is beautiful music, no instruments are needed in the meetings.

———

A Nazarene brother who received the baptism with the Holy Ghost in his own home in family worship, in trying to tell about it, said, "It was a baptism of love. Such abounding love! Such compassion seemed to almost kill me with its sweetness! People do not know what they are doing when they stand out against it. The devil never gave me a swet thing, he was always trying to get me to censuring people. This baptism fills us with divine love."

———

The gift of languages is given with the commission, "Go ye into all the world and preach the Gospel to every creature." The Lord has given languages to the unlearned Greek, Latin, Hebrew, French, German, Italian, Chinese, Japanese, Zulu and languages of Africa, Hindu and Bengali and dialects of India, Chippewa and other languages of the Indians, Esquimaux, the deaf mute language and, in fact the Holy Ghost speaks all the languages of the world through His children.

———

A minister says that God showed him twenty years ago that the divine plan for missionaries was that they might receive the gift of tongues either before going to the foreign field or on the way. It should be a sign to the heathen that the message is of God. The gift of tongues can only be used as the Spirit gives utterance. It cannot be learned like the native tongue, but the Lord takes control of the organs of speech at will. It is emphatically, God's message.

———

During a meeting at Monrovia, a preacher who at one time had been used of God in the Pentecost Bands under Vivian Dake, but had cooled off, was reclaimed, sanctified and filled with the Holy Ghost. When the power of God came on him his eight-year-old son was kneeling behind him. The boy had previously sought and obtained a clear heart, and when the Holy Ghost fell on his father, He also fell on him and his hands began to shake and he sang in tongues.

———

Bro. Campbell, a Nazarene brother, 83 years of age, who has been for 53 years serving the Lord, received the baptism with the Holy Ghost and gift of tongues in his own home. His son, who was a physician, was called and came to see if he was sick, but found him only happy in the Lord. Not only old men and old women, but boys and girls, are receiving their Pentecost. Viola Price, a little orphan colored girl eight years of age, has received the gift of tongues.

———

Mrs. Lucy F. Farrow, God's anointed handmaid, who came some four months ago from Houston, Texas, to Los Angeles, bringing the full Gospel, and whom God has greatly used as she laid her hands on many who have received the Pentecost and the gift of tongues, has now returned to Houston, en route to Norfolk, Va. This is her old home which she left as a girl, being sold into slavery in the south. The Lord, she feels, is now calling her back. Sister Farrow, Bro. W. J. Seymour and Bro. J. A. Warren were the three that the Lord sent from Houston as messengers of the full gospel.

The inaugural issue of the *Apostolic Faith*, the official newspaper of the Azusa Street Mission.

the Holy Spirit, he was not sure if he was worthy of such a gift. He wrote:

> The first day in the meeting I sat to myself, away from those that went with me. I began to thank God in my heart for all things, for when I heard some speak in tongues, I knew it was right though I did not understand it. Nevertheless it was sweet to me. I also thanked God for Elder Seymour who came and preached a wonderful sermon. His words were sweet and powerful and it seems that I hear them now while writing. When he closed his sermon, he said, "All those who want to be sanctified or baptized with the Holy Ghost, go to the upper room; and those that want to be justified come to the altar." I said that is the place for me, for it may be that I am not converted and if not, God knows it and can convert me.

That evening Mason had a dream in which he believed God was telling him not to try so hard to get the Spirit. If he could only allow God to do the work, the blessing would surely be his. The next day Mason was prepared to empty himself in order to be filled with the Holy Spirit.

> I got a place at the altar and began to thank God. After that, I said Lord if I could baptize myself, I would do so; for I wanted the baptism so bad that I did not know what to do. I said, Lord, You will have to do the work for me; so I turned it over into His hands....Then, I began to seek for the baptism of the Holy Ghost according to Acts 2:44 which readeth thus: "Then they that gladly received His word were baptized." Then I saw that I had a right to be glad and not sad.

Some said, "Let us sing." I arose and the first song that came to me was "He brought me out of the Miry Clay." The spirit came upon the saints and upon me.... Then I gave up for the Lord to have His way within me. So there came a wave of Glory into me and all of my being was filled with the Glory of the Lord. So when He had gotten me straight on my feet, there came a light which enveloped my entire being above the brightness of the sun. When I opened my mouth to say "Glory," a flame touched my tongue which ran down to me. My language changed and no word could I speak in my own tongue. Oh! I was filled with the Glory of the Lord. My soul was then satisfied.

For the next five weeks Mason and his companions praised God, shouted with joy and thanksgiving, and opened themselves to the power of the Holy Spirit.

By the time they left the city and the revival in May, all three had received the gift of speaking in tongues, and they returned to Memphis "saved, baptized and Holy Ghost filled." Jones was leery of the incorporation of Pentecostal experiences into Holiness worship. Naturally he and Mason clashed about whether to introduce the doctrine into the Church of God in Christ. Jeter and Young also were split over the issue, even though both of them had experienced the indwelling of the Spirit while in California. Jeter joined forces with Jones, while Young sided with Mason. By the time their general assembly met in Jackson, Mississippi, that year, the Pentecostal controversy had thoroughly engulfed the denomination. With Jones and Jeter leading the charge, the assembly voted to withdraw the right hand of fellowship from Mason and Young. In response, the two men led fourteen congregations out of the assembly into an alternative convention. In subsequent years Mason took Jones to court over the use of the name Church of God in Christ. Because Mason had incorporated the organization, the court ruled he had legal right to the name.

Even as the Church of God in Christ and the Church of Christ, Holiness, went their separate ways, they continued to share a motivation to renew black religion with the power of slave religion. As blacks moved into urban centers in both the South and the North, the Christianity of their ancestors would provide the continuity necessary to ease the transition. In cities like Chicago, Detroit, New Orleans, and New York, Holiness and Pentecostal doctrine would be preached, taught, moaned, and sung in storefronts and tent meetings that harked back to a time when life was much simpler, when the land yielded enough food to sustain a family, when Christians went to church to sing God's praises and not their own, and when the redemptive power of Christ's death and resurrection was enough to sustain the believer from day to day. In the city, even though one's neighbor might be only two yards away rather than two miles, life could be cold and lonely. For many new arrivals anything from home was good, and religion above all else was the way they soothed the psychological scars left by racial discrimination and residential segregation. Yet for many others not even the Good News of the churches could repair the damage done by urban living. Some blacks found city life so disorienting that they felt alienated from everyone—whites, other blacks, and even their God. In the midst of this confusion they sought refuge in new gods, new religions, new ways of making sense of their new realities.

The early-twentieth-century figure Father Divine represents the best and the worst in black people's perennial flight for religious, political, and social freedom in America. In his own way, although it was dangerously parochial, Father Divine attempted to carve out a religious institution focused on improving the lives of America's disinherited. Through strict teachings on positive thinking, divine healing, and pious living, Father Divine envisioned leading countless Americans from different racial and economic backgrounds to a new homeland where peace, happiness, and wealth flourished for all.

Father Divine's early life remains fiercely debated, however, recent scholarship locates his origins in Rockville, Maryland, where he was born in May 1879 to George and Nancy Baker, former slaves. George Baker, Jr., as he was named, wrestled daily with poverty, eating a diet that consisted mostly of cornmeal and pork fat and working odd jobs to help support the family.

Soon after his mother died in 1897, George left Rockville searching for a place promising better employment and social opportunities for African Americans. His journey landed him first in Baltimore. In storefront churches in the city, he received mentoring from a number of ministers. His faith deepened, and the idea of developing his own ministry received nourishment. At this time, George also plunged himself into readings from the New Thought movement, including religious bodies such as the Unity School of Christianity and Christian Scientology. Their teachings on spiritual prosperity, positive thinking, and pious living ignited a fire in George's soul.

His new interest led him to move in 1906 to the so-called center of the New Thought movement—California. He traveled from Los Angeles to San Francisco and talked with many followers and ministers in the movement. But while in California, George found himself drawn to another religious movement, the Azusa Street Revival in Los Angeles. George, too, received the baptism of the Holy Ghost and spoke in tongues. In fact, his experience increased the already simmering fire in his soul.

George soon returned to Baltimore, where he joined a small Baptist congregation, Eden Street Church. One Sunday morning in 1907, George met Samuel Morris, who claimed to be the Father Eternal. George was intrigued by Morris's revelation that he was God. The two embarked on a ministry etched in New Thought theology. George changed his name to Messenger, God in the Sonship. As their ministry blossomed, the Messenger, with his captivating messages, rose in prominence. He soon began disputing with his colleagues over the direction of the ministry, and in 1912 left to build his own vision.

The Messenger's primary interest was preaching to the poor the way to acquiring heaven on earth. He traveled through parts of Georgia, garnering a small but dedicated following—most of them women. His New Thought teachings, however, created a stir among established African American clergy. After run-ins with clergy and policemen in several small Georgia cities, the Messenger headed north.

The Messenger, who had married a woman named Peninniah from Macon, Georgia, before leaving the South, opened up shop in Brooklyn in 1917. He believed he could secure a following if he planted himself away from the city and places like Harlem, where he felt prominent black churches and social and political activity would pose too many obstacles to his securing members. Immediately after moving to Brooklyn, he changed his name to Major Jealous Divine. His followers called him Father Divine and his wife Mother Divine. His charismatic appeal stemmed in part from the astute manner in which he merged seemingly disparate religious traditions—such as Roman Catholicism, Pentecostalism, New Thought (similar to Scientology), and Methodism— into his own constructed theology. As in his operations in the South, Father Divine provided housing, food, and counseling to those in need.

In 1919 Father Divine purchased a home in Sayville, a Long Island community of middle-class whites. He and his wife became the town's first black homeowners. He opened his home to his followers and created an employment agency that provided domestic jobs in the area. Moreover, here he could hone his teachings and practices. In the evenings he and his followers held Holy Communion, a feast that included a table of finely prepared food. It ended with singing, shouting, and speaking in tongues. While Father Divine offered great resources to his followers, he also required them to share their earnings, adhere to celibacy (though he was married), refrain from drinking alcohol, relinquish their ties to their families, and adopt new names. Father Divine kept his home and land finely manicured. His followers dressed impeccably, and he wore the finest suits available. In 1930 as many as twenty-six adults and four children lived in Sayville with Father Divine.

His white neighbors were often outraged by what appeared to them to be ostentatious behavior, such as owning a fancy car and clothes. Just as important, they refused to accept an African American in their town. Many white homeowners began an effort to drive him out. After he had been taken to court on several occasions and accused of disturbing the peace (his Holy Communion services would often last late into the evening), Father Divine received jail time for violating neighborhood codes. Soon after, in 1932, he moved his headquarters to Harlem. He established his Peace Mission ventures that same year, and opened restaurants around New York and in Newark, New Jersey. His mission was to provide a clean space in which people could receive home-cooked meals at a low price. Soon his followers, known as angels, began opening restaurants, hotels, and other businesses around the country. The Peace Mission Movement attracted both African Americans and whites. It is estimated that Father Divine's movement had as many as 10,000 members during its heyday and nearly 150 Peace Mission centers nationwide.

In 1943 Mother Divine died, and Father Divine married the white Canadian Edna Rose Ritchings three years later. Their marriage was reportedly "purely spiritual."

Father Divine (1880–1965) was a controversial minister in Harlem.

The explosive growth of Father Divine's organization created problems. It meant not only supervising a larger pool of people but also attracting increased attention from outsiders. Investigations by journalists and government officials led to allegations of homosexuality, sexual abuse among women, and the mishandling of funds being reported by several newspapers. The allegations pointed to Father Divine as well as to other workers all over the country.

Father Divine was a tragic figure. His efforts to free himself as well as others from economic and spiritual enslavement ended up tainting his own vocation as well as the lives of many others. Today Father Divine is often maligned for his outrageous theological claims, and many debate his accomplishments in ameliorating the harsh social conditions facing African Americans, but what remains uncontested is his ingenuity in persuading people to give him their money and join his movement. Historians and theologians may argue over his tactics and ideology, but his social programming, albeit brief, provided relief and hope for many hopeless women and men. Father Divine died in 1965, though his followers believe he is still alive.

7

Black Gods of the City

I said, "You are gods, and all of you are children of
the Most High. But you shall die like men, and fall like
one of the princes." Arise, O God, and judge the earth;
for You shall inherit all nations.

PSALMS 82:6–8

Throughout the first half of the twentieth century, African Americans remained an overwhelmingly Southern people. The 1910 U.S. Census reported that roughly 90 percent of blacks lived in the South, and only about two out of every ten lived in cities. But blacks flocked to urban areas at an incredible rate between the two world wars. Some sought higher wages promised in industrial plants in the Midwest and Northeast, and as news of the good life up North trickled back to Southern rural areas, more and more blacks decided to cast caution to the wind. For countless others falling cotton prices, increased mechanization of farming, and the tyranny of the sharecropping system left no choice but to leave the land of their birth for parts unknown. Even more simply sought refuge from the lynch mobs and separate and unequal schools, housing, and justice. They believed things had to

Part of Jacob Lawrence's acclaimed Migration series.
This painting is called *The Migration Gained in Momentum.*

If You are a Stranger in the City

If you want a job   If you want a place to live
If you are having trouble with your employer
If you want information or advice of any kind

CALL UPON

The CHICAGO LEAGUE ON URBAN CONDITIONS AMONG NEGROES

3719 South State Street

Telephone Douglas 9098   T. ARNOLD HILL, Executive Secretary

No charges—no fees. We want to help YOU

A card distributed by the Chicago Urban League
offering help to African American migrants.

be better in the North, where promises of a new start, a new identity, and a new deal took on the tone of salvation.

So by the thousands they went—to Chicago, Gary, Detroit, New York, Boston, and wherever else they could find a place to be somebody. The males arrived first. They wanted to find good jobs and good homes before sending for the rest of their family. But by the time the Great Migration subsided, it seemed as if entire communities were moving. From 1910 to 1920 Chicago's black population increased from about 44,000 to nearly 110,000, while during the same period Cleveland's black community swelled from roughly 9,000 to close to 35,000. The number of blacks in Detroit began at about 6,000 in 1910, increased to 41,000 in 1920, then skyrocketed to 120,000 in 1930. They packed their steamer trunks with all their earthly possessions, headed to the closest rail station, and bought one-way tickets to the end of the line. Among the many things black migrants took with them was their faith in God. If they were going to survive in the big city, they would need God as their foundation. Besides, God had brought them out of slavery, and it would be up to God to lead them to the Promised Land.

Just as they had in the rural South, the migrants turned to the church to fill all their spiritual, social, and economic needs. But they often found the established churches in their new hometowns inhospitable. One can easily see how the daily arrival of hundreds of new black residents from the Deep South might upset the balance of community life in the urban North. In the past their relatively small numbers had posed no threat to whites. But suddenly even blacks who had come in small numbers earlier came to fear the presence of more and more black country folk. They worried that the presence of so many unsophisticated newcomers would set back the progress they had worked so hard to create. The only way to ensure the continuation of good relations with whites, they reasoned, was to socialize the migrants in the ways of the city and encourage them to drop at once all the backward habits they had brought with them. This negative attitude toward the black migrants was just as prevalent among black churches as it was in the white world.

While many of the new urban dwellers chose to adapt to religious life in the city for the sake of social mobility, others found it difficult and unnecessary to give up their old traditions. Baptists, Methodists, and the like set up churches in storefronts where they could be free to sing the spirituals that had nurtured their foreparents through slavery and shout God's praises until their bodies gave way. Many of them were attracted to the Holiness Pentecostal piety, which did not restrict the movement of the Holy Spirit as they thought some of the mainstream denominations did. Yet for each one who looked to the church for answers to the many riddles encountered in urban bureaucracies and institutions, there was one who decided that the God of

The outside of a storefront church in Chicago, photographed by Russell Lee in 1941.

Christianity had run his course. Why did it seem that life was no better in the North than it had been on the plantation? Where was the Christian God in July 1919, when 38 black people were killed and over 1,000 left homeless after rioting whites attacked Chicago's black residents? Did God still hear black voices as they called out for help? Did God still care about all their woes, about the injustices perpetrated in God's name?

Easter morning services at a storefront church in Chicago.

Some blacks concluded that they did not get the answers they needed because they were petitioning the wrong god. What more could one expect from the god who also heard the prayers of the lynch mobs in Mississippi? Maybe the answer was to get a new black god who understood what black folk needed and was always on their side. Maybe they needed a religion that placed them at the top of the social order—a little closer to God than those who committed evil

Noble Drew Ali, founder of the Moorish Science Temple of America.
Born Timothy Drew in 1886, he founded the new religion based on the guiding
principles of several faith traditions, including Christianity and Islam.

against blacks in God's name. In 1913 Noble Drew Ali provided such a religion.

Born Timothy Drew in 1886 in North Carolina, Noble Drew Ali arrived in Newark, New Jersey, mysteriously. No one is sure of the details of his life before his emergence as a religious leader in the North, but various legends point to a man who had unique beginnings. One myth claims that he was an orphaned child of slaves in North Carolina. With no surviving family to rear him, Timothy was nurtured by Cherokee Indians. Another legend has him being raised by a band of gypsies until one gypsy woman noticed his spiritual nature and sharp intellect. She took him with her to Egypt, where he studied the ancient mystery cults of northern Africa. A third tale has Drew studying Islam in Morocco and Saudi Arabia, where religious leaders gave him their certification and blessing to teach Islam in America. All the legends about his origins were designed to do the same thing: convince black and white Americans that Noble Drew Ali's religious authority did not come from whites.

Newark's black community accounted for less than 3 percent of the city's population in 1913, when Ali established his Canaanite Temple there. But New Jersey was well on its way to earning its reputation as "the Georgia of the North" because of its high incidence of racial violence. Newark proved to be fertile ground for Ali's religious philosophy of racial superiority, based on the premise that the oppression and degradation blacks experienced in the United States was the result of their lack of knowledge about their true identity. They were not Negroid at all; they were Moors, or Moorish Americans, descendants of Moroccans brought to America. Because Ali believed they were an Asiatic people, the so-called Negroes' true religion was Islam, not the white man's Christianity. Ali urged his followers to wear fezzes and turbans, to carry identification cards that signified their racial heritage

Members of the Moorish Science Temple at an annual gathering in 1928.

as Moors, and to adopt a flag resembling Morocco's—signs of their rediscovery of their true selves.

Ali's teaching was important and distinctive in several ways. First, it moved blacks away from the racial labels they had been assigned by whites. Negro, nigger, and colored were not accepted any longer. Second, it disassociated black Americans from the negative images of black

Africa, even though it did not deny the reality of sub-Saharan primitivism. Most important, Ali gave black Americans a new god, Allah, to call their own. By claiming Islam as the true religion of Moorish Americans, Ali provided a new religious outlet for blacks that did not overlook the color consciousness of American religion, as Christianity often did. Instead, it took responsibility for blacks' slow social progress out of white hands and placed it squarely on black shoulders. This transfer gave blacks an opportunity to control their own destiny instead of ceding their power to whites.

Within a decade membership in the organization grew to an estimated 30,000, with branch temples in several cities, including New York, Detroit, Chicago, and Philadelphia. As the movement spread, Ali was unable to maintain control, and splinter groups began to form. In 1923 internal power struggles at the Newark temple led a group of members to form the Holy Moabite Temple of the World. Frustrated by the fracture in the Newark community, Ali moved his organization's headquarters to Chicago that year and renamed it the Moorish Holy Temple of Science. The movement continued to grow, particularly in communities receptive to the Pan-African message of Marcus Garvey's Universal Negro Improvement Association. Even though Garvey's organization had a religious component, the African Orthodox Church, he did not compel members of the UNIA to hold membership in the church for fear of alienating potential recruits. As a result, Ali found an audience among Garveyites. Once Garvey was convicted on charges of mail fraud in 1925, many of his disillusioned followers flocked to Ali and his band of Moors.

In 1927 Ali wrote the *Holy Koran,* a catechism for the instruction of new converts. In it he combined information from the Quran and the Bible, as well as *The Aquarian Gospels of Jesus Christ* and *Unto Thee I Grant.* Ali made certain to account for everything—the origins

The origins of the African Orthodox Church in the United States can be traced to the philosophical and religious teachings of the Universal Negro Improvement Association, the African American political organization founded by Marcus Garvey in 1914. At one point the largest independent African American political association in America, the UNIA had strong roots in what would today be termed Afrocentric religious philosophy. Christian language and symbols could be found throughout the organization's mission statement, newspaper, and hierarchical structure, and in Garvey's lectures and speeches. The UNIA, although it championed the rights of African Americans, encouraged members to believe in the common humanity of all women and men. Its teachings claimed that Jesus was black and that his resurrection symbolized the mark of liberation for all oppressed and "crucified" black peoples. The Harlem-based UNIA held meetings on Sundays, during which prayers, singing, and sermons were commonplace. The organization had chaplains at the local and national levels, who were under the supervision of the High Executive Council, the UNIA's governing body. The association's ethos encouraged individuals to believe in themselves and maintain faith in God and in the black peoples.

Without a doubt Garvey established a political movement rooted in Christian themes and philosophy. However, some scholars argue that Garvey and his organization had no ties to the church's creation. In fact, some question whether Garvey ever officially joined the church. This camp believes the African Orthodox Church was the work of George Alexander McGuire alone.

Recent scholarship suggests that McGuire acquired a burning desire to create an independent black religious body based on his ties to the UNIA. In fact, in several letters and sermons he acknowledges the UNIA as providing the impetus for the church's founding.

The beginnings of the African Orthodox Church date to 1919, when McGuire returned to the United States from Antigua and joined the UNIA. McGuire, an Anglican priest, had grown tired of the Episcopal Church's snail's pace response to complaints about the small number of black leaders in the church. In Harlem, McGuire found a new home in the UNIA. The organization's emphasis on black pride and nationalism provided the tools for McGuire to begin reconstructing the role he believed religion should play in the African American struggle for liberation. The teachings of the UNIA led him to draft *The Universal Negro Catechism* and *The Universal Negro Ritual,* and in 1919 he founded the Independent Episcopal Church of the Good Shepherd, an African American congregation, in New York City.

From this congregation emerged the African Orthodox Church. Reportedly McGuire and Garvey disagreed on the role the church would play in the UNIA, which influenced McGuire to relinquish his position as the UNIA chaplain-general. Garvey believed his organization was not designed to encourage its followers to support any one religious body. But within a year Garvey and McGuire reconciled, though the African Orthodox Church and the UNIA had no official ties. Nevertheless, the ministers and members of the African Orthodox Church regularly participated in UNIA events and even held leadership positions in the group.

The African Orthodox Church borrowed its liturgy from Catholicism and the Anglican Church, but its message of black nationalism and autonomy resulted in a religious body with a radically different theology and interpretation of the Bible and its teachings.

of the universe, the differences between the races, the true nature and heritage of so-called Negroes, and the lineage of Jesus. His question-and-answer format was designed to make it easier for readers to memorize large chunks of information. Ali described how Moors in America came to be known as blacks:

> Name some of the marks that were put upon the MOORS of Northwest, by the European nations in 1774? Negro, Black, Colored and Ethiopian. Negro, a name given to a river in West Africa by MOORS, because it contains black water.
>
> What is meant by the word Black? Black according to science means death.
>
> What does the word colored mean? Colored means anything that has been painted, stained, varnished or dyed.
>
> What does Ethiopia mean? Ethiopia means something divided.

Can a man be a Negro, Black, Colored or Ethiopian? No.

Why? Because man is made in the image and after the likeness of God, Allah.

He showed how the labels given to blacks by whites were meant to keep them ignorant of the fact that they were God's children.

Ali knew that his version of Islam could not simply invalidate Christianity, for complete repudiation of a religion blacks knew so well would make it difficult for them to hear the truth of his message. Instead, he incorporated and reinterpreted elements of the Christian epic—most important, Jesus' genealogy—to fit his racialized version of Islam.

Who was Jesus? He was a prophet of Allah.

Who were His father and mother? Joseph and Mary.

Will you give in brief the line [genealogy] through which Jesus came? Some of the Great Fathers through which Jesus came are: Abraham, Boaz by Ruth, Jesse, King David, Solomon, Hezekiah and Joseph and Mary.

Why did Allah send Jesus to this earth? To save the Israelites from the iron-hand oppression of the pale-skin nations of Europe, who were governing a portion of Palestine at that time.

What was the nationality of Ruth? Ruth was a Moabitess.

What is the modern name for the Moabites? Moroccans.

Ali recast Jesus as a Moorish African revolutionary prophet sent by God to free his people from the treachery of Europeans. Like Jesus, Ali was a Moorish prophet sent to end the oppression of his people. His exercise in selective genealogy read the so-called Negroes back into the family of God, locating them within both Jewish and Christian traditions while placing both religions in the racial milieu of northern Africa.

In 1928 Ali changed the name of his organization to the Moorish Science Temple of America. The movement was growing at such an alarming rate that he appointed a group to help him oversee the various temples. He tried to pick men with significant formal education in the hope that their learning would benefit the rank-and-file membership. Unfortunately, he discovered that several of his appointees were fleecing the poor temple members, demanding they purchase religious amulets and other trinkets.

A couple of the rival leaders tried to split the Moorish Science Temple by discrediting Ali's status as a prophet of Allah. Apparently Ali was romantically involved with three women, two of whom he married in Moorish wedding ceremonies. While he would argue that polygamy was in line with Quranic and biblical teaching, the fact that his wives were fourteen and sixteen years old was much more difficult to rationalize. Moreover, the pregnancy of his youngest bride put him in violation of Illinois's statutory rape laws. After fifteen years of prosperity and growth, the Moorish Science Temple of America was in jeopardy of collapsing.

In 1929, while Ali was on business in another city, one of his rivals, Sheik Claude Greene, was assassinated. The police issued a warrant for Ali's arrest, and he was detained as soon as he returned to Chicago. He proclaimed his innocence and eagerly awaited trial so he could prove that he had no connection to the murder. Unfortunately, that opportunity never came. On July 20, 1929, three weeks after being released on bond, Noble Drew Ali died. Some say he was murdered by Greene's followers, while others believe he was killed by members of the Chicago Police Department. Either way, it is clear that Ali's death left his Islamic movement in chaos. However, Ali had instructed his followers many times that he would never leave them alone—after his death he would return to them, reincarnated in a new leader for a new age.

As mysteriously as Noble Drew Ali appeared in Newark, W. D. Fard materialized in 1930 in Detroit. He claimed to have been born in Mecca, Saudi Arabia, to the tribe of Kumiesh, the same tribe as the prophet Muhammad's. Fard also told followers he'd earned degrees from Oxford University in England and the University of California. As his popularity grew and claims of his divinity spread, detractors would insist that he was nothing more than a small-time hustler who found his big score among naïve and spiritually hungry migrants from the rural South.

Master Fard Muhammad, founder of the Nation of Islam. Believed to be Allah by his followers and the true savior of black people, he selected Elijah Muhammad as his messenger on earth.

Detroit's automobile industry was a magnet for black workers in search of higher wages, but it quickly became apparent that good jobs came at great cost. Economic downturns are often first felt among auto producers, and as the nation sank into the Great Depression blacks in Detroit were among the first to feel the pain. In 1930 nearly 50 percent of the black male workforce was concentrated in the automobile sector, with almost 40 percent of black women workers in domestic services. As factories laid off employees and white women made the decision to clean their own homes, black families in Detroit were left with few economic options. Moreover, racial hostility increased as whites viewed recently arrived blacks as competition for the remaining jobs, decent housing, and public assistance.

In this context W. D. Fard was able to gather a following among black newcomers, former Garveyites, and disaffected members of the Moorish Science Temple. He was slight of build with a very light beige complexion, which often led him to be mistaken for a white man. He peddled silks, oils, and other exotic products that he claimed were from Africa and the Far East. Working his way from door to door through Detroit's black community, Fard took every opportunity to include in his sales pitch information about the black man's natural homeland. As soon as his host expressed a desire to know more, Fard would suggest that he come back. He would dine with the family, eating whatever food they were able to scrape together for their mysterious guest, but after dinner the "lessons" would begin. "Now don't eat this food," he told one dinner host in 1930. "It is poison for you. The people in your own country do not eat it, since they eat the right kind of food. Since they do not eat it they have the best health all the time. If you would live like the people in your home country, you would never be sick anymore."

Of course, according to Fard, diet was only one part of blacks' problems. More poisonous than the pork they had been fed for centuries was their unhealthy reliance on Christianity, the white man's religion. Islam, he claimed, was the natural religion of the black race in America. And, like Noble Drew Ali, Fard argued that once blacks came to a proper knowledge of themselves, they would begin to enjoy the good health and prosperity Allah intended for them. Among the many people who responded to Fard's message was Elijah Poole. A thin, frail man who migrated to Detroit from Georgia with his family in search of better-paying jobs, Poole was at the end of his rope when he heard the truth according to Fard.

Born Robert Poole in Sandersville, Georgia, in 1897, he'd spent six years working in a Chevrolet plant in Detroit until the Great Depression forced him out of his job. Emasculated by his inability to find steady work and provide a decent living for his wife and eight children, he sought refuge in whiskey. The Pooles survived on government assistance for two years while Elijah drowned his sorrows in self-pity and liquor. Then one day his wife told him about Fard, who was speaking that evening at Liberty Hall, the former headquarters of the local chapter of the Universal Negro Improvement Association. After some cajoling Poole agreed to attend the lecture.

The place was so packed it was difficult for Poole to find a seat. Disillusioned with organized religion and with God, he wondered what all the fuss was about—he couldn't imagine Fard having anything to add to the countless other religious entrepreneurs who preyed on black folk. From spiritualists and mediums who offered contact with the dead to Pentecostal preachers who offered the healing power of Jesus, Poole had seen it all. He was underwhelmed by Fard's physical appearance. What could this white-looking man possibly say that would make his life better? But once Fard began to speak, Poole became convinced that there was something different about this one, something that would change the history of black people in America.

In a rousing speech about the condition of the race, Fard told the crowd that they needed to reject the labels given them by whites. He contended that the opposite of what they had been led to believe about themselves was true; it was the black man, the so-called Negro, who was made in the image of God, Allah; and whites were simply envious of black skin, since they had lost all their color. While Europeans were still living in caves and wallowing in their own feces, the black man had grand civilizations that would put the godless American culture to shame. Like Noble Drew Ali, Fard said the so-called Negro descended from Afro-Asiatic roots. Blacks were stolen members of the tribe of Shabazz in Mecca and not from the primitive cultures of sub-Saharan Africa. He said that he was there "to find and bring back to life his long lost brethren, from whom the Caucasians had taken away their language, their nation, and their religion."

Fard also had a convenient explanation for the origins of the white race. He taught his followers that an evil black man who had wanted all the glory of Allah for himself had created whites by permitting only lighter-complexioned blacks to mate. Over thousands of years the process had resulted in a race of people that had lost not only their melanin but also the ability to do good and be in the right relationship with Allah. Fard told the crowd about a new organization he had established, the Nation of Islam, which would teach the so-called Negro all he or she needed to know about the true order of the universe and the black man's role in it. One of his lessons taught:

> The original people must regain their religion, which is Islam, their language, which is Arabic, and their culture, which is astronomy and higher mathematics, especially calculus. They must live according to the law of Allah, avoiding all meat of "poison animals," hogs, ducks, geese, possums, and catfish. They must give up completely the use of stimulants, especially liquor. They must clean themselves up—both their bodies and their houses. If in this way they obeyed Allah, he would take them back to the Paradise from which they had been stolen— the Holy City of Mecca.

The onus was on blacks to do something for themselves instead of waiting for whites to get an attack of conscience or for the white man's God to answer the prayers of the righteous black man.

Fard's rationale for the condition of the black race in America seemed quite plausible to Poole. Moreover, he concluded that such a keen insight into the nature of humanity could only have come from on high. As Fard walked through the crowd after his address, Poole approached him. "I know who you are," he told the mystery man. "You are God himself." Fard replied, "That's right, but don't tell it now. It is not yet time for me to be known." Poole joined the Nation of Islam and studied with Fard. Like Noble Drew Ali before him, Fard urged members to rid themselves of the slave names given them by whites and to reclaim their true Muslim names. He gave Poole the name Elijah Karriem; later Poole became Elijah Muhammad. For three years he worked hand in hand with Fard and eventually was named chief minister of the Nation of Islam. Then, in 1934, Fard seemingly vanished, never to be heard from again.

Theories abound regarding Fard's disappearance. Some suggest he was run out of town by the police, who feared the uncontrollable growth of a nationalist organization like the Nation of Islam. At one point even Elijah Muhammad said that Fard had been deported to Saudi Arabia. Another theory suggests that he moved back to his old stomping ground in California, where he and a recently paroled partner in crime got reacquainted and ran new scams and hustles in the Bay Area. The Nation of Islam teaches that Fard, now remembered as the Honorable Fard Muhammad, was caught up and taken away by a giant spaceship. From his vantage point on high he waits for the proper time to return and bring wrath upon the wicked Caucasians and those people of color who refuse to hear the message of Allah.

Elijah Muhammad helped build the Nation of Islam, a religious organization advocating the moral, economic, and political improvement of blacks.

Once Fard left Detroit, a power struggle ensued. Elijah Muhammad believed that, as Fard's right-hand man, he should succeed him. Rivals believed otherwise, and as the competition for control heated up, Elijah Muhammad fled Detroit and established the Nation of Islam Temple 2 in Chicago. From that location he ran a branch of the Nation that eventually outlasted its rivals in Detroit.

Elijah Muhammad quickly added a couple of nuances to the organization's doctrine. Most important, he established the deification of Fard Muhammad. By making Fard into Allah in the flesh, Elijah Muhammad solidified his own place as Messenger of Allah, since he had been second in command while Allah walked the earth. Naturally many blacks and whites alike found this new twist to Nation of Islam theology farcical at best. Local and federal law enforcement agencies, as well as black religious leaders, began to circulate information about Fard's criminal record in California. Others criticized Elijah Muhammad's background, suggesting that his personal problems made him the wrong person to lead a mass religious movement in America.

While deification of a leader may have seemed odd to outsiders, it certainly was not unique to the Nation of Islam. Neither was the movement's emphasis on the African origins of civilization and world religions. By 1931 George Baker, known to the world as Father Divine,

BLACK JEWS

In 1990 the Council of Jewish Federations estimated that 2.4 percent of self-identified Jews, about 132,000, reported their race as black. Though the council acknowledged it had not distinguished in the survey the individuals identifying as African Americans, officials estimated the African American Jewish population at 100,000. While the number of African American Jews in the United States remains tiny in comparison to the number of African American Protestants, Catholics, and Muslims, their

Rabbi Arnold Josiah Ford led a congregation of black Jews in Harlem during the Great Depression.

journey adds another dimension to the portrait of African American history. African American Jews, in their interpretation of the Hebrew Bible, share similar and significant bonds with the Black Liberation Theology movement, which emerged from African American Protestant traditions. Indeed, like their Protestant sisters and brothers, African American Jews saw themselves in the stories of the Old Testament, but unlike black Protestants, African American Jews (in large numbers) saw their lineage stretching back to the Ethiopian Jews. In fact, several early movements in the African American Jewish community envisioned blacks as the *original* Jews. Of course, their claims of African ancestry within Judaism raised concerns among white Jews. Nonetheless, the rise and development of African American Jews illuminates the extent to which African Americans retrieved their own history and configured it with what was handed them in the New World. The story of African American Jews is about neither their legitimacy nor the accuracy of their claims to a Jewish lineage. It is a story of a people creating a community held together by traditions and beliefs that reinforce their humanity in a society where they are often ignored and maligned.

Early scholarship suggests that the black interest in Judaism stems, in part, from three points: During slavery Africans saw themselves as (literally) the Hebrews of the Old Testament. African American leaders such as Booker T. Washington regularly encouraged African Americans to emulate Jews. Finally, several African Americans saw Jewish practices and beliefs as directly rooted in African history, specifically that of

Ethiopia. Recent scholarship indicates that blacks arriving in the United States from the West Indies and South America may have possessed Jewish ancestry. Indeed, since the seventeenth century Jews from countries such as Portugal and Spain had been establishing communities in several parts of Latin America and the Caribbean. All these factors and more contributed to the development of African American Jews.

Signs of African American Jewish communities and congregations in the United States began surfacing at the beginning of the twentieth century. The largest concentrated period of growth of African American Jewish congregations occurred between 1908 and 1925. Not surprisingly, the rise of African American Jewish congregations happened alongside the Great Migration, when thousands of African Americans settled in big cities in the North. Some of the more popular and larger African American Jewish congregations rooted themselves in places such as Philadelphia, New York City, and Washington, D.C. Cities such as Chicago and Newark also witnessed the arrival of sizable numbers of African American Jews. But in Philadelphia, New York City, and Washington, D.C., black Jews drew attention as leaders in the East Coast Civil Rights movement. Many African American rabbis in particular were visible and vocal protest leaders, especially in Marcus Garvey's Universal Negro Improvement Association.

From 1919 to 1931 at least eight African American Jewish congregations emerged in Harlem. During that period Harlem was also home to a large number of white Jewish immigrants. Though the two groups did not relate to each other in significant ways, a handful from each community shared information and interacted in worship services.

One of the most important figures in African American Jewish history is Rabbi Arnold Josiah Ford. Born and reared in Barbados, Ford arrived in New York in the mid-1900s. The jazz musician and composer dived into the African American cultural scene in Harlem soon after his arrival. One of his major musical accomplishments was his composition of the Universal Ethiopian Hymnal for Garvey's UNIA. Alongside his interests in music, Ford possessed deep religious roots. In 1923 he was named rabbi at the Beth B'nai Abraham (House of the Sons of Abraham), an African American Jewish congregation in Harlem. Fluent in Hebrew and Yiddish, Ford was prominent among African American Jews in the New York area. Rabbi Ford, closely aligned with Marcus Garvey, dreamed of a Jewish homeland for African Americans in Africa. Proclaiming African Americans' connection with Ethiopians, Ford immersed himself in Ethiopian politics and visited the country several times. In fact, he took a delegation of temple members with him to Ethiopia between 1931 and 1932 in the hope of securing land for an African American settlement. His dream never materialized, though, and he died in 1934.

After Rabbi Ford's death, Rabbi Wentworth A. Matthew appeared on the scene. Rabbi Matthew, who had worked side by side with Rabbi Ford, was a major personality and made the congregation much more visible. He spoke vehemently about blacks' role in history in general and in Jewish history in particular. Here are his words from one memorable sermon:

> During slavery they took away our name, language, religion, and science, as these were the only possessions the slaves had, and they were pumped full of Christianity to make them more docile.... All so-called Negroes are the lost sheep of the House of Israel which can be proved from scripture and they all have birthmarks that identify their tribe. Jacob was a black man because he had smooth skin.

> Rabbi Matthew had about 1,000 congregants, most of whom were West Indian.
>
> In Washington, D.C., Bishop H. Z. Plummer led a Jewish congregation of the Church of God and Saints of Christ and claimed to have been the "impersonation" of "Grand-Father Abraham." In Philadelphia the Jewish congregation was called the Church of God, and its leader was Prophet Cherry, who believed that blacks were the "true" Jews.

had established his Peace Mission Movement in Harlem and promoted himself among his followers as God. Likewise, forms of Judeo-centric Christianity such as Prophet Cherry's Church of God, and William S. Crowdy's Church of God and Saints of Christ, as well as black Hebrew movements like Rabbi Wentworth A. Matthew's Beth B'nai Abraham in Harlem, taught black followers in American cities that blacks, not Europeans, were the chosen people of Yahweh.

The Nation of Islam under both Elijah Muhammad and Louis Farrakhan has been thought of primarily as a political organization, and there are a number of good reasons for this interpretation. First, as a black religious nationalist organization, it has mixed religion and politics. For example, Elijah Muhammad consistently opposed enlistment in the United States armed forces and even encouraged members to refuse induction if drafted during times of war. At the same time, he was not a pacifist. His objection to fighting for America blurred the line between secular and sacred by claiming that blacks who bore arms for the United States contributed to the work of the Devil. Members of the Nation of Islam were free to fight for sacred causes, defined as those that directly benefited the proliferation of Islam as the true religion of the black race. So a traditionally political matter like draft evasion was given a religious meaning.

Second, rarely is nation building thought of in religious terms in the American context. Not a single nation in the western hemisphere looks at itself as a theocracy. Yet with the Nation of Islam it was impossible to distinguish calls for a racial state from Elijah Muhammad's understanding of the role of Islam in that state. He wrote:

> We cannot be successful in the house of our enemies; we should be in our own house. That which is other than our own is for those who are other than our own. "Our own" is unlimited physically and spiritually. There are those who think our lack of freedom, justice and equality can be solved in the white man's crooked and corrupt politics.... I have many times said that the solution to our problem is divine. There are so many who would—just for self-praise or exaltation—like to lead you astray under the false claim that they can solve the problem by ways other than the divine. You should never listen to these leaders because they will lead you into the fisherman's net. Such leaders show no respect for Allah and His power to solve our problem

of freeing us from our enemies and raising us into a state of independence like other independent nations.

He called for the creation of an independent nation-state for black Americans as a religious solution to the race problem, not a political one.

The influence of groups like the Nation of Islam and the Moorish Science Temple has always exceeded their relatively small membership. Nonetheless, African American openness to religious alternatives signified a deep frustration with the pace at which Christianity responded to the oppression of blacks. At every major juncture in American history, the nation held out to blacks the promise of full participation in the republic in exchange for loyal service and commitment to the country's well-being. And each time America awarded black steadfastness with unfulfilled agreements and occasionally open hostility. Over 400,000 blacks served in segregated units of America's armed forces during World War I. Yet instead of receiving a heroes' welcome home, black soldiers in Spartanburg, South Carolina, were beaten by local whites for seeking service at a segregated café. In both world wars countless thousands of blacks heeded the call to fill urban factory positions vacated by whites sent off to combat. But instead of thanks they received death threats from white coworkers and physical abuse from those who did not want blacks living near them.

Perhaps Christianity was the wrong religion for blacks; it certainly did not seem to benefit them in the ways many thought it should. As increasing numbers of blacks opted for new forms of religious expression and fulfillment, mainstream religious organizations sought ways to respond to them. But mainstream black churches suddenly found themselves under pressure from a different direction. The Civil Rights movement was bringing the political struggle for equal rights from the streets into the church.

8

Prayers of the Righteous

Let not the 12 million Negroes be ashamed of the fact
that they are the grandchildren of the slaves. There is no
dishonor in being slaves. There is dishonor in being
slave-owners. But let us not think of honor or dishonor
in connection with the past. Let us realize that the future
is with those who would be truthful, pure and loving.
For, as the old wise men have said, truth ever is,
untruth never was. Love alone binds. And truth and love
accrue only to the truly humble.

—MAHATMA GANDHI

Simplified stories of the African American Civil Rights movement
often begin and end with Martin Luther King, Jr. During his thirty-
nine-year life, King played a crucial role in the struggle for black equality.
But he was part of a movement. It is important to recall that many men
and women of God had long been challenging on a mass scale racial segregation in America. African American encounters with Gandhian nonviolence stretch back a generation before the Montgomery bus boycott.
King learned the tactics of boycotts and sit-in protests from black people
who studied the parallels between black and Indian oppression long

Dr. Martin Luther King, Jr., in 1961.

before Rosa Parks took her fateful ride. The truth is that King and other civil rights activists were inheritors of a grand tradition of nonviolent direct action—one that found support and encouragement in Gandhi's use of faith to stir social action against oppression.

The line from Mahatma Gandhi to Martin Luther King, Jr., stretches through a young black Baptist minister named Howard Thurman. On a trip to India with students from Howard University, where he was dean of the chapel, Thurman had an encounter with Gandhi that led to an epiphany. The meeting of the Baptist and the Hindu leader of India's fight against British colonialism had been arranged as the highlight of a trip Thurman led to South Asia in 1935. He planned stops in India, Burma, and Ceylon. Before they left Howard's campus in Washington, D.C., Thurman knew that the one man they wanted to meet was Gandhi. Thurman was initially underwhelmed as he sat on a straw mat next to the elderly man. Gandhi was a small, emaciated man. The hunger strikes he used to pressure the British to respect the rights of Indians had left their mark on him. And when he spoke, he spoke softly. There were none of the strutting, grand gestures, and loud demands for respect and attention that Thurman was accustomed to seeing in American leaders.

But the spirit, a rush of energy, flowed from Gandhi. The little man engaged Thurman and the others with an intense gaze that met and held Thurman's stare. In their first minutes together, Gandhi presented Thurman a simple gift, a piece of cotton woven on Gandhi's own spinning wheel. As Thurman ran his fingers over the smooth fibers, Gandhi reminded his visitors that Indian cotton represented India's drive to be self-sufficient and its protests against British imperialism and economic exploitation. And with his little gift Gandhi gently mentioned how cotton tied black Americans and Indians together. Thurman suddenly grasped that Gandhi had selected this gift with care. This little Indian man knew that in the United States blacks had picked cotton as slaves and their stolen labor had been the basis for the white South's wealth. On behalf of blacks all across America, Thurman accepted the gift as a token of solidarity in the struggle for full equality.

The idea behind the trip was to let the young people of South Asia meet black American Christians and spread the word of Christ through firsthand accounts of black life and religion in America. But Thurman, a commanding speaker, went silent as Gandhi talked with rare power. It came to Thurman that Gandhi spoke as the Christian prophets once spoke. God-given faith, Gandhi explained, is a tool for revolution against the might of British guns and, possibly, the oppression of white American segregation. He began to press Thurman to talk. What did he think of Indians mustering spiritual power to challenge the arsenals of the British Army with nonviolence? Just as Thurman was taking hold of the revelation that Gandhi was opening for him, Gandhi pushed him to a yet higher level that made Thurman uncomfortable. Why are blacks in the United States so committed to Christianity? Gandhi asked. The seemingly wispy old man challenged Christianity as a religion used by whites to keep blacks in place with images of a white Christ and promises of rewards in heaven for slaves who obeyed and even loved their masters. Christian faith, Gandhi argued, seemed to lack the spiritual strength to be the backbone of a black American movement for liberation. Hindu principles had offered a basis for nonviolent opposition to British power. Did Christianity have similar power to fuel an uprising of oppressed American blacks?

In his conversation with Gandhi, Thurman's world of Christian faith was turned inside out and upside down. His very soul was twisted and prodded by Gandhi's words, and his Christian life in America suddenly seemed hollow. Segregated Christian churches in his hometown of Daytona, Florida, did little to condemn racism and nothing to bring blacks and whites together, even in the pews. When the grandson of slaves attended Morehouse College in the early 1920s, he had seen segregation at the YMCA, the Young Men's Christian Association. At every turn Christian brotherhood in American churches came second to the need to keep the races separate and to protect white privilege. When Thurman went north to study at Rochester Theological Seminary, he quickly learned that America's race problem was not limited to the South. In 1927 seminary officials reprimanded Thurman, along with two white classmates, because they chose to share a dorm room. Up to that point the seminary had admitted two black students per class so they could share living quarters and avoid integration. On another occasion Thurman was asked to speak on behalf of the student body at a memorial service for the school's white custodian. But before he could, administrators requested that he withdraw from the program to avoid embarrassing the man's family.

These thoughts flooded into Thurman's mind as he listened to Gandhi. The strange little man at whom he had initially stared out of curiosity now held his attention as one worthy of admiration. The meeting ended with laughter and hugs, but in his heart and soul Thurman was in turmoil. His very life in the Christian church was under challenge. He began to ask those with him if they thought Christianity was "impotent to deal radically, and therefore effectively, with the issues of discrimination and injustice on the basis of race, religion, and national origin." Had Christianity in the United States been perverted by bigots? Or was it a religion of the powerful and the white? At times moved to tears, Thurman later wrote that he had a personal epiphany that clarified for him what God required of true Christians.

Near the end of our journey we spent a day in Khyber Pass on the border of the northwest frontier. It was an experience of vision. We stood looking at a distance into Afghanistan, while to our right, and close at hand, passed a long camel train bringing goods and ideas to the bazaars of North India. Here was the gateway through which mogul conquerors had come in other days bringing with them goods, new concepts, and the violence of armed might. All that we had seen and felt in India seemed to be brought miraculously into focus. We saw clearly what we must do somehow when we returned to America. We knew that we must test whether a religious fellowship could be developed in America that was capable of cutting across all racial barriers, with a carry-over into the common life, a fellowship that would alter the behavior patterns of those involved. It became imperative now to find out if experiences of spiritual unity among people could be more compelling than the experiences which divide them.

Finding "spiritual unity" across racial barriers did not require Thurman to abandon Christianity. His faith in Christ was great. And the African American allegiance to Christianity was far too deep for him to imagine converting substantial numbers of people to a new religion. To him faith in Jesus Christ as God's son and humanity's savior had nothing to do with racism. The bigotry in the Christian Church had to do with powerful people hiding their greed and abuse of their neighbors behind distortions of God's words. Thurman prayed for some way to prove that blacks' commitment to Jesus was not in vain.

Emboldened by what he would later call his Khyber Pass experience, Thurman returned home determined to demonstrate to himself and America—indeed, to the world—that Christianity had the power to promote racial equality through understanding and common ground based on shared faith. To him this newly vital Christian faith that could power a Civil Rights movement required fellowship among all people, regardless of race or class. Over time he planned to journey beyond the bounds of the Sunday church service and affect the way people of different backgrounds interacted in their daily lives. Gandhi's techniques of appealing to the conscience of the world and nonviolent resistance to British troops had created a revolution in India. An American movement for racial justice would take a strong commitment on the part of believers, both black and white, to transcend a history of pain and find his conception of life's treasure, a core faith in God. Thurman began preaching this message when he returned to the United States. In order to make the journey to faith, he said, black believers would have to leave the cocoonlike safety of all-black churches and spread their wings in search of a new spiritual harbor in racial inclusiveness, a place of equality and integration.

Between 1936 and 1939, Thurman canvassed the United States, speaking to church groups, student organizations at colleges and universities, ministers' meetings, summer camps, and any other organization that would listen to his message. His trip to Asia had been sponsored by the YMCA and YWCA, with grants from the World Christian Student Fund. Most groups he visited on his return simply wanted to hear about his travels in exotic lands, and about the health of Christianity in a region dominated by Hindus, Muslims, Buddhists, and Sikhs. They wanted confirmation of Christianity's superiority and its ability to save heathens from eternal damnation. But Thurman spoke about Christianity in a new light, as a beacon of justice shining bright on the sinful racial and cultural divide in America. His lectures about his Khyber Pass experience provided the moral and intellectual energy for his widely read book *Jesus and the Disinherited.* Although it was published ten years after his trip, the book was inspired by his encounter with his brown brothers and sisters in South Asia. Thurman's writings broke ground for black Christians and were highly optimistic about the possibilities for brotherhood in Christ's name in America.

> Churches have been established for the underprivileged, for the weak, for the poor, on the theory that they prefer to be among themselves. Churches have been established for the Chinese, the Japanese, the Korean, the Mexican, the Filipino, the Italian, and the Negro, with the same theory in mind. The result is that in the one place in which normal, free contacts might be most naturally established—in which the relations of the individual to his God should take priority over conditions of class, race, power, status, wealth, or the like—this place is one of the chief instruments for guaranteeing barriers.

Thurman's proposal was to transform the church from a cornerstone of the status quo into a Gandhian-style base for social struggle that carried the banner of racial justice in the name of Jesus. This goal required white and black Christians in the United States to see and acknowledge that American Christianity itself was deeply flawed in the 1940s. Thurman took to pulpits nationwide to deliver this difficult message. He said the teachings of the Savior, Jesus, had been distorted by the sin of racism. He said America's history of persecution, conquest, and cultural chauvinism had led believers far afield from the humble, universal teachings of the man from Galilee. The key to solving the church's race problem resided in getting back to the simple core values of Jesus. That meant, according to Thurman, loving your brother, sister, and neighbor without regard to race. The idea was simple: Racism is against Christ's teachings and out of line with being a Christian. The only way to walk in Christ's path was to see beyond skin color and condemn anyone fixated on race. This truth applied to white people and also to black people. Thurman asked people in every church to free their minds of the need to see the color of a person's skin before they saw the person as fully human and a representation of the living Christ.

He wrote:

> The religion of Jesus says to the disinherited: "Love your enemy. Take the initiative in seeking ways by which you can have the experience of a common sharing of mutual worth and value. It may be hazardous, but you must do it." For the Negro it means he must see the individual white man in the context of a common humanity. The fact that a particular individual is white, and therefore may be regarded in some over-all sense as the racial enemy, must be faced; and opportunity must be provided, found, or created for freeing such an individual from his "white necessity." From this point on, the relationship becomes like any other primary one.

It wasn't long before Thurman and others who sought an end to American Christianity's racial barriers understood that it would take many years for the walls of racial separation to come tumbling down, if they ever did. The revelation hit in May 1939, at a uniting conference of the various branches of Methodism held in Kansas City, Missouri. The meeting was called to hammer out a plan of action for bringing together black Christians across denominations to fight segregation. The two largest groups represented were the Methodist Episcopal Church, North, and the Methodist Episcopal Church, South. In 1844 slavery and abolitionism had fractured their common cup, and since then regional social customs and negative attitudes about each other had kept these denominations apart. The uniting conference promised to erase nearly a century of animosity in one affirmative vote.

The only missing elements were the African Methodist denominations, who were thinking about uniting into a single African Methodist Church.

While the white Methodists of the North and South found much to agree about, one thing still perplexed their plans for unity: what to do about the 340,000 black Methodists who were members of the Methodist church in the North? Southern Methodism had basically rid itself of its "Negro problem" in 1870 by helping its former black members establish the Colored Methodist Episcopal Church, and Southern Methodists were not interested in worshiping with blacks, sending their children to church camps with black children, or having blacks in positions of authority in their regional conferences. Black Methodists, by contrast, were determined to fight for full participation in the united body, regardless of whether Southerners were offended by their presence.

White Methodists from the Northern church held the pivotal position. They refused to support any type of agreement that expelled blacks from the church, yet they were not willing to risk unification over matters of social custom. As planned, the conference concluded with the formation of the United Methodist Church, but its unity existed in name only. While the church was divided by region into annual conferences, all black members of the United Methodist Church were herded into one huge unit called the Central Jurisdiction. The new church welcomed Jim Crow to the membership roles and passed by the chance to reclaim the religion of Jesus—a caste-free fellowship.

Despite the racist assumptions behind the formation of the Central Jurisdiction, blacks in the United Methodist Church were able to maintain their churches and attracted large numbers of young people. The key young leader in the Central Jurisdiction congregation was James Morris Lawson. Born in Ohio in 1928, Lawson had learned of Gandhi while in college and become a strong supporter of the principles of nonviolence. Gandhian philosophy had deepened Lawson's allegiance to the idea that there was power in the willingness to suffer and sacrifice for a cause. Even after Gandhi was assassinated on January 30, 1948, Lawson was unwavering in his commitment to using Gandhi's tactics to fight segregation in America. To Lawson's mind, Gandhi's movement went beyond political battle; it was a way of life that included a spiritual disavowal of violence. Lawson acted on Thurman's and Gandhi's messages in 1951, when the U.S. Army drafted him to serve in the Korean War. He refused. Serving in an army that killed people troubled his conscience, he said, because it violated his understanding that all conflict could be resolved through nonviolence.

Refusing the draft carried the possibility of going to jail unless the draft board recognized him as a conscientious objector. The prospect of being jailed certainly frightened Lawson, but he told friends his spirit was buoyed by the belief that Jesus would have made the same choices. Christianity, he believed, affirms all human life and meets the enemy with the open arms of love instead of the clenched fists of hate. As he awaited word on his case, he stood firm in the faith that God would provide a way out of his situation. Later that year Lawson was sentenced to three years in federal prison. After serving only thirteen months he was released on the condition that he would do church work in India.

After his parole in 1952, Lawson was ordained in the United Methodist Church. He spent three years teaching and serving as campus minister at Hislop College in Nagpur, India. While Lawson was technically a Methodist missionary, he used the opportunity to study Gandhian nonviolence. His battle to stay out of prison had prepared his mind for the possibility of using nonviolence not just to resist oppression but to call attention to poverty, racism, failed schools, and other ills that often go without comment in modern American life.

When Lawson returned to the United States in 1956, racial tensions had the nation on edge. In 1954 the Supreme Court had ordered

James Lawson, a United Methodist minister,
taught nonviolent resistance in the Civil Rights movement.

the desegregation of America's public schools. That decision prompted massive resistance from white Southern politicians, many of whom had been elected on the promise of keeping their states segregated. After the Supreme Court's ruling, however, the federal government and the courts stood against the segregationist politicians and with black and white people working for racial justice. Buoyed by the government's support for integration, younger black people stepped up their demands for an immediate end to segregation. On April 30, 1957, blacks in Greensboro, North Carolina, voted to boycott local movie theaters after the Reverend Melvin Swann was forced to sit in the Jim Crow balcony during a showing of *The Ten Commandments.* And from December 1955 through December 1957 a young, previously unknown Baptist minister named Martin Luther King, Jr., led blacks in Montgomery, Alabama, in a citywide bus boycott that drew international press coverage. King's supporters chose to walk—day after day, rain or shine—rather than continue to ride the city's segregated buses.

King was not the first black minister at Dexter Avenue Baptist Church to take on God's armor in the battle for a righteous cause. The church's previous pastor, the Reverend Vernon Johns, was a dynamic and well-known speaker in his own right, but his commitment to social activism in a town like Montgomery brought the church all sorts of unwanted attention. Unlike most black and white ministers in the evangelical South, Johns did not condone "literal" interpretations of the Bible. Often he would preach sermons from texts that lent themselves to racial interpretations, and he never passed up an opportunity to mock the behavior of white Christians. He did not believe their actions exemplified the religion of Jesus, no matter what they said.

Even Johns's own congregation was unable to avoid his frequent barbs. Dexter Avenue was one of the elite black congregations in Montgomery, and while Johns did not fault them for being successful, he constantly reminded them that the teachings of Jesus emphasized social responsibility. It was not enough to have individual achievement; Johns felt the good news of Christ required the disavowal of all social distinctions, including class. By 1952 the membership of Dexter Avenue had grown weary of their pastor's sharp critiques. Perhaps he did not know when to stop pushing, or maybe his punches landed a bit too close to home. Johns was finally forced to resign and move back to his family farm outside Petersburg, Virginia.

The twenty-eight-year-old Martin Luther King, Jr. Soon after he moved to Montgomery, Alabama, he helped lead the Montgomery bus boycott.

Within a year of his departure, Martin Luther King, Jr., was called to Dexter Avenue Baptist Church. He was a surprise choice to lead a major church in the state's capital. He was twenty-four years old and newly wed to a classical singer when he arrived in September 1953 from Boston, where he was still a graduate student at Boston University. A short and dashing fellow who always seemed to be dressed in his Sunday best, King was best known as the son of Martin Luther King, Sr., a famed preacher in Atlanta. Young King was fortunate to get the Dexter Avenue pulpit, but he took the job as a stepping-stone. He hoped to get back to Atlanta or some other big city after a few years in Montgomery. He saw Dexter Avenue as a place where he could focus on completing his doctoral dissertation and start a family. Between home, church, and school, he had no time for local politics, and little time for new friends, such as the Reverend Ralph Abernathy.

King had only the most fleeting awareness of the budding Civil Rights movement. He had read Howard Thurman's writings and heard him speak, but King was far from enlisting in an army of nonviolence. One thing he knew though, from watching his father through the years, was that the call to the pastorate was a call to service. There would be nights when his work at the church would prevent him from spending quality time at home, and there would be days when crises in the congregation would keep him from his scheduled writing time. With that in mind he tackled his work at Dexter Avenue, fully prepared to take on whatever burdens God saw fit to lay at his feet.

In fact, the ministry had become somewhat of a King family business. In 1894 King's maternal grandfather, the Reverend A. D. Williams, was elected pastor of Ebenezer Baptist Church in Atlanta, Georgia. Williams built Ebenezer into one of the more notable black Baptist congregations in Atlanta, along with churches like Wheat Street and Friendship. When he died of a heart attack on March 21, 1931, Michael L. King, Sr., later known as Martin Luther King, Sr., was selected to succeed him.

Most likely the younger King learned a great deal about the work of the church from hearing stories about his grandfather. He also saw his father's example of taking the faith beyond the church. The senior King was very active in religious, civic, and political affairs. By the time the younger King was in the fourth grade, his father had organized a protest against segregated elevators in the county courthouse, worked on voter registration drives with the local chapter of the NAACP, and been elected a member of the executive committee of the Atlanta Civic and Political League. When Martin Luther King, Jr., went to the

The Reverend Ralph Abernathy, King's confidant throughout the Civil Rights
movement, and Bayard Rustin, a behind-the-scenes strategist
for King and the movement, flank King in 1956.

all-male Morehouse College in Atlanta between 1944 and 1948, he saw
his father's example reinforced by a series of black men who talked
about leadership.

Those who lectured frequently at Morehouse included Howard
Thurman; the NAACP's Walter White; the editor and writer W. E. B.

Du Bois; Mordecai Wyatt Johnson, president of Howard University; and Channing Tobias, executive secretary of the colored work division of the YMCA. But those speeches were secondary to the weekly chapel talks of Morehouse's president, Benjamin Mays. The tall, dignified Mays gave talks that were legendary for their fire and passion. He would challenge the students to sharpen their intellectual and critical skills in order to prepare for life's tests. It was as if every Morehouse man was being groomed for greatness. It was as if they were being prepared for a time when wisdom, courage, and immovable faith would be required of them. For Martin Luther King, Jr., the Montgomery bus boycott was a time to test his soul.

When Rosa Parks was arrested for refusing to surrender her seat in the whites-only section of a city bus, King had no idea what was to come. Mrs. Parks said her decision was "a matter of dignity—I could not have faced myself and my people if I had moved." Eventually, local leaders such as Jo Ann Robinson, head of a women's group that had long protested segregated buses, and E. D. Nixon, the former leader of the local chapter of the NAACP, embraced Mrs. Parks's cause. They brought the movement to King's church and enlisted the young, charismatic minister in their march for justice. King recalled, "We discovered that we had never really smothered our self-respect and that we could not be at one with ourselves without asserting it."

What was originally planned as a one-day bus boycott was quickly growing into an ongoing protest. And although he was just twenty-six years old, and may have had doubts about being ready for his moment as a leader, King found that his ability as a speaker, his education, and his family's deep connections among religious leaders in the South made him ideal to head the boycott movement. If the choice had been solely his, King most likely would have declined any invitation to lead the movement in Montgomery. If the nonviolent boycott failed, his leadership would bear the brunt of the responsibility. This new pastor ran the risk of alienating a congregation and a community he was just getting to know. And if the boycott was a resounding success there was also risk; he would expose himself and his family to the reprisals of local white segregationists. But his sense of duty led him to follow the will of God.

With the official motto "Justice without Violence," the Montgomery Improvement Association under the direction of the Reverend Martin Luther King, Jr., embarked on a "walk for freedom" that changed the course of American history. With love as their only weapon, over 90 percent of black bus patrons in Montgomery chose to walk the pavement rather than to continue to participate in the system of American apartheid. Several newspapers drew a comparison between the boycott and Mahatma Gandhi's strategies in fighting British colonialism. And, like that of Gandhi, King's public role in the movement drew legal challenges as well as violent responses. Not only was he arrested for leading a boycott white officials deemed illegal but on January 30, 1956, his home was bombed. King was not at home at the time, but one can only imagine the rage he felt when he was informed of the incident. King did all he could to suppress his instinct to strike back with violence, because he knew the nation, both whites and blacks, was watching to see how he would respond. Instead of demanding an eye for an eye, he found the spirit to rise above his human anger and urge peace and love in the tradition of lectures he had heard from Howard Thurman about a thin, passionate Indian leader named Gandhi. King said:

> This is a spiritual movement and we intend to keep these things
> in the forefront. We know that violence will defeat our purpose.
> We know that in our struggle in America and in our specific

struggle here in Montgomery, violence will not only be impractical but immoral. We are outnumbered; we do not have access
to the instruments of violence. Even more than that, not only is
violence impractical, but it is immoral; for it is my firm conviction that to seek to retaliate with violence does nothing but
intensify the existence of evil and hate in the universe. Along
the way of life, someone must have sense enough and morality
enough to cut off the chain of hate and evil. The greatest way
to do that is through love. I believe firmly that love is a transforming power that can lift a whole community to new horizons
of fair play, good will and justice.

So, with songs of faith and praise on their lips, the love of God
in their hearts, and their minds "stayed on freedom," the black community of Montgomery boycotted bus transportation from December
1955 until segregated seating on the buses was ruled unconstitutional
in January 1958.

While the bus boycott was under way in the American South,
James Lawson was in South Asia reading about King's emerging leadership in Indian newspapers. When Lawson returned to the United
States in 1956, he enrolled at Oberlin School of Theology in Ohio to
complete his ministerial training. By 1957, however, he could no
longer sit on the sidelines. The power of Jesus, he later said, was with
the people standing against the evil of segregation, so he dropped out
of seminary and took a job with the Fellowship of Reconciliation, an
international pacifist organization. He opened an FOR field office in
Nashville, Tennessee, where he also enrolled at Vanderbilt Divinity
School. Part of Lawson's responsibilities with the group was to train
volunteers in Gandhian tactics of nonviolent direct action. He taught
students how to organize sit-ins, pray-ins, wade-ins, and any other form

There is no doubt that music has long played a key role in black life, both in the Civil Rights movement and in the religious and social life of blacks in general. During times of slavery music was a means of expression both mournful and joyful. Robbed of the ability to communicate in many other ways, slaves raised their voices in songs that took form from African tonal patterns modified by the European religious songs heard in their masters' churches. Music sustained the slaves as they worked, and it united them in their lives. Reconstruction saw the rise of independent African American churches, which in turn led to blacks' composing and publishing their own music. A hundred years later, during the Civil Rights movement, music again became prominent as gospel came out of the churches and into the streets to embolden and unite protesters.

By all accounts Lucie E. Campbell was a pioneer in gospel music. Before Thomas Dorsey merged sacred themes with secular music to give the world the gospel blues, Campbell combined the slow pace of Baptist lined hymn traditions with European classical music to give us the gospel waltz. Her contribution to black sacred music makes her one of the giants of the African American religious experience.

Born just outside Duck Hill, Mississippi, in 1885, Campbell experienced the highs and lows of life almost from the beginning. Her mother gave birth to her on the train after a visit with Lucie's father, a railroad worker. When he received word that his ninth child had arrived, he rushed home. But he was killed in a train accident and was never able to see his new child. Through her early years Lucie experienced indifference from her mother, perhaps related to feelings of guilt and hostility about Burell Campbell's death. In the end, though, it was Lucie of all the children who became closest to her mother and took care of her.

When Lucie was just two the family moved to Memphis in search of better wages and opportunities. Lucie, a bright child, taught herself to play the piano by watching her older sister's lessons from outside the room. After Lora completed practice, Lucie tried to replicate the teacher's instructions. Some days she would get it right, but most of the time she ended up devising some other method that made sense to her. In addition to being a quick study in music, she mastered Latin and other subjects in school, becoming valedictorian of her high school class. Although she would not complete her B.A. degree from Rust College until 1927, education was always her love, and she spent nearly five decades as a public school teacher in Memphis.

Lucie Campbell's music career began in earnest in 1916, with her election as music director of the National Baptist Training Union. The organization was one of the major departments of the National Baptist Convention, designed to help mold and maintain black Baptist identity. Campbell was responsible for making sure local units had music that created the proper spirit in worship. She also had to ensure that the musical selections at the annual meeting represented sound Baptist theology. It was a strange time to be a member of the National

Baptist Convention. The organization had experienced a major schism in 1915, and no one was certain of its future. In that context music became more important than ever. It set the stage for worship, but it also helped galvanize people into a more unified body. Miss Lucie, as she was called by those who worked with her, assumed a more prominent role as the convention became aware of her talents. By the time she attended her last convention, in 1962, all the annual meeting's musical selections had to be approved by her, and participants had to audition for singing slots at the weeklong session.

In 1919 Campbell introduced the National Baptist Convention to a woman she believed was a rising star. She beamed with pride as Marian Anderson enthralled the delegates with her rich tones and graceful presence. That same year Campbell published two songs: "The Lord Is My Shepherd" and "Something Within," which she composed after hearing a blind boy on Beale Street tell a heckler that something within prevented him from singing the blues; she later invited the boy to perform it at the convention. It was a resounding success and, though adapted to contemporary music, is still a favorite in African American churches today. In 1921 Campbell participated in the development of *Gospel Pearls*, an African American songbook published by the Sunday School Publishing Board of the National Baptist Convention. This landmark publication

included hymns and early gospel songs by pioneering black composers like Campbell and the Reverend Charles Albert Tindley. While Campbell published a significant amount of original material from 1919 through the late 1930s, she always relinquished copyright to either the National Baptist Training Union or the National Baptist Young People's Union. She believed it was improper for church musicians to seek to profit from the gifts God freely gave them.

As musical director of the National Baptist Training Union Congress for forty-three years, and because the National Baptist Convention was the largest organization of African Americans, Campbell stood as the gatekeeper of black sacred music for nearly half a century. Year after year delegates from across the country heard music she selected, then took the new songs back to their congregations and choirs. Lucie E. Campbell left an indelible mark on the black religious experience.

Mahalia Jackson was born October 26, 1911, in New Orleans, the third of six children. Her father worked as a barber and preacher, and her mother died at age twenty-four, when Mahalia was just four. After her mother's death Mahalia was raised by two maternal aunts. Church was a big part of her life while she was growing up, and it is where she gained attention as a singer. As a youngster she joined the children's choir at Plymouth Rock Baptist Church in New Orleans and became well versed in the musical traditions of black Baptists.

While her Baptist roots played a major role in the development of her vocal style, Jackson also was influenced by the

sanctified church next to her home. Most of all she was impressed by the full-bodied style of spiritual expression that most Baptists would have regarded as excessive. Sanctified worshipers used instruments often associated with secular music, like drums and guitars, and they clapped their hands and danced. This unbridled passion for the Lord made its way into Jackson's style; she was occasionally criticized by good Baptists for being "of the world." But she continued to sing praises to God with her entire body. Another important ingredient in Jackson's gospel vocals was blues music. Ma Rainey and Bessie Smith had a tremendous impact on her style, and while she remained faithful to her roots in sacred music, she recognized that both gospel and blues were responses to the black experience in America.

Jackson moved with an aunt to Chicago in 1927 and gained immediate attention because of her unique sound. She joined the Greater Salem Baptist Church and after the first rehearsal was made the choir's lead soloist. Jackson also joined the Johnson Singers, a female a cappella group, and won major parts in local gospel plays. All the storefront churches in Chicago knew about her deep-throated, raw yet holy sound, but Jackson wanted more people to hear the message of her music.

Her singing career moved up another level in 1935, when she became the official song demonstrator for Thomas Dorsey. Known today as the father of gospel music, Dorsey was a prolific composer and publisher, and he needed soloists to give his music a public face. He had also been the pianist for Ma Rainey, so Jackson's bluesy style was a perfect fit for his musical temperament. She remained a demonstrator until the close of 1945, when she signed a recording contract with the Apollo label. Although she had been recording since 1934, Apollo put her on the map in black communities across the nation. Her 1946 recording of "Move on Up a Little Higher" earned her one of the first two gold records ever for gospel music. Off the popularity of that song, she was appointed the official soloist of the National Baptist Convention. Part of her responsibility was to attend state and national meetings and sing at major functions as the representative of the convention.

While all of this gained Jackson national attention with black audiences, she still wasn't as well known in white households. That began to change in 1954, when she agreed to a one-year deal to host a weekly gospel program on CBS Radio. That same year she switched to the Columbia recording label and almost instantly became a crossover success. In 1958 she appeared on *The Ed Sullivan Show* and *The Dinah Shore Show*. Not only were white Americans aware of her music now but Europeans had also discovered her passion. Even though they often could not understand the words, they could feel the pathos and the hopefulness in her music.

As the Civil Rights movement unfolded in the 1950s and '60s, Jackson lent her support to leaders she believed were "for real." She accompanied Martin Luther King, Jr., on countless fund-raising rallies and sang "I've Been 'Buked and I've Been Scorned" at the March on Washington just before his now famous "I Have a Dream" speech. When King died in 1968, Jackson sang his favorite song, "Precious Lord, Take My Hand," at the funeral. She died on January 27, 1972, of heart failure at the age of sixty, and Aretha Franklin closed her funeral with a soul-stirring rendition of "Precious Lord."

of action that would create pressure on the nation's legal but immoral segregation.

The idea, as Lawson interpreted it to black and white students who began to flock to his training sessions from across the nation, was to push racism into the open through nonviolent protest. He wanted to force the American people to confront the absurdity of it and struggle through it to the eventual point of love, respect, and understanding. Lawson put all his faith in a combination of nonviolent resistance and the power of Christ's example for true believers to endure suffering to defeat evil. He told his trainees:

James Lawson presiding over a meeting in 1960
in Nashville concerning lunch counter integration.

When you are a child of God...you try thereby to imitate
Jesus, in the midst of evil. Which means, if someone slaps you
on the one cheek, you turn the other cheek, which is an act of
resistance. It means that you do not only love your neighbor
but you recognize that even the enemy has a spark of God in
them, has been made in the image of God and therefore needs
to be treated as you, yourself, want to be treated. Jesus is very
clear about this: "Do unto others as you want others to do unto
you." Not as they do unto you, but as you *want* them to do unto
you—which is a rather powerful ethic for personal relation-

The Reverend Wyatt Tee Walker, a leader in the movement with
Dr. King, at a restaurant in Petersburg, Virginia, in 1960.

ships, regardless of whether family or school or community or
nation.

Lawson's Nashville office of the FOR began working with every
civil rights group in the nation, including the newly created Southern
Christian Leadership Conference, headed by the hero of the Montgomery bus boycott, Martin Luther King, Jr. But Lawson developed a
particularly close alliance with the young people in the Student Nonviolent Coordinating Committee and helped them launch a sit-in
movement in Nashville and several other Southern communities

through the first half of the 1960s. Lawson's acolytes integrated everything from public swimming pools to waiting rooms in bus terminals.

For Lawson and John Lewis, a seminary student and SNCC leader who later became a congressman from Georgia, nonviolence was more than a political tactic. It was the only way Christians could actively engage in struggle without violating the humanity of another human being. Moreover, they believed the Civil Rights movement was a miracle of faith. It was a moment in history when God saw fit to call America back from the depths of moral depravity and onto his path of righteousness. The blot of racial discrimination was out of step with God's will, they argued, and the suffering of blacks provided a vehicle through which America could save its soul. In Lawson's opinion, the movement "judged America; it said to America, Repent of your violence, repent of your militarism, repent of your systemic violence, repent of your hatred of all people, not just black people."

The road to nonviolence was not always an easy one. Lawson and other movement leaders had to contend with both the violent reactions of white segregationists and the skepticism of blacks who saw it as foolish and suicidal. "Teaching non-violence in the '50s was a major challenge," said Lawson, "because it was like teaching a foreign language, even though it was a language deeply rooted in the spirituality of Jesus, deeply rooted in the spirituality of many of the prophetic stories of the Hebrew Bible." The challenge, then, was to convince black people in the South that nonviolence was rooted in their own history and religion. "We had to try to show people that here was a magnificent history that was a secret that we were trying to [unpack]," exclaimed Lawson; "that King was not a man from Mars, but a man out of the black church and out of the black Scriptures." More often than not they were able to achieve their goal of spreading the gospel of nonviolence. And like David, who stood against Goliath and the forces of

Dr. King and other civil rights leaders in Atlanta in 1960.

domination and repression he represented, black Christians resisted evil by turning the other cheek and loving the enemy into submission.

While Lawson and Lewis were motivated by a deep spiritual commitment to nonviolence and integration, other black activists openly ridiculed their tactics. Malcolm X, the national spokesman of the Nation of Islam, said black people had been beaten and killed by segregationist violence for too long. The Muslim minister called for self-defense and separation from whites instead of integration. Additionally, he called into question the logic of the love ethic upon which nonviolent direct action was built. He proclaimed on many occasions, "It is not possible to love a man whose chief purpose in life is to humiliate you and still be considered a normal human being." The test for the veracity of Malcolm's critique of the Civil Rights movement, nonviolence, integration, and the love ethic would come in the summer of 1966, during James Meredith's one-man Walk Against Fear. Meredith was walking to encourage Southern blacks to take the risk of registering to vote despite segregationists' intimidation.

Meredith was the first black graduate of the University of Mississippi. He had gone to school there even though the governor had personally blocked the door to the registrar's office in an effort to bar his entry. Now, with the same persistence, Meredith set out on a sixteen-day march across Alabama and Mississippi to protest the use of terrorist violence to control the black population, and on the second day he was gunned down by a white man.

Badly wounded, Meredith could not continue the walk. Vowing not to allow the evil of racism to impede their journey to full equality, Martin Luther King, Jr., and other civil rights leaders met in Mississippi to finish the march in his stead. While all agreed that to stop the march would be a victory for segregationists, not all agreed about whether to continue to advocate integration and nonviolence. Once

Stokely Carmichael issued his cry for black power, the Christian energy behind nonviolence seemed old and tired. College students who had been drawn to Lawson's teaching about nonviolence now found themselves taken with the defiant attitude of Malcolm X and the Nation of Islam in asserting the moral superiority of blacks over the white devils.

How would Christianity respond to black power? At its foundation, the promise of black power was the intoxicating vision of a once enslaved people finally standing tall, strong, and independent. By this standard, talk about loving segregationists looked weak and nonviolent strategies, impotent. Gandhi's teaching of nonviolent protest rooted in faith, as interpreted by Thurman, Lawson, and King, held the hearts and minds of most black Americans. But black power created new tensions for black religious leaders as young, impatient black activists pushed for immediate answers that nonviolence and even the deepest faith seemed hard-pressed to deliver.

9

A Call to Witness

Is your all on the altar of sacrifice laid?
Your heart, does the Spirit control?
You can only be blest and have peace and sweet rest,
As you yield Him your body and soul.
—"IS YOUR ALL ON THE ALTAR?"

At its heart, the Bible's good news is that God so loved mankind that he willingly sacrificed his only son to save lost souls from sin and death. And that son, Jesus Christ, knew he had to suffer and die to fulfill his father's plan. Jesus told the apostles that his purpose on earth was to "give his life as a ransom for many." In contemporary America the most popular understanding of a martyr is a fearless warrior willing to give his or her life for a cause. However, early Christian notions of martyrdom did not require death. Anyone who picked up Jesus' cross to speak the truth and fight for righteousness stood as a Christ-like martyr.

The Civil Rights movement of the 1950s and '60s had many martyrs; most are not known. They never got their names in the newspapers, and no monuments have been built in their honor. Why did so

The Freedom Choir of the Tabernacle Baptist Church sang inspirational songs at civil rights meetings in Selma. Religious music that could galvanize protesters was an important part of the movement's formula for success.

many people, including children, walk tall in the face of death threats, snarling dogs, and racist attacks? One key to black Americans' willingness to sacrifice is that every Sunday the gospel told them God would not desert them in their time of trial if they had the courage to walk with Christ. Ordinary people went to jail, lost jobs, and risked injury for what they saw as Christ-like suffering for the greater good. In their minds they held a common vision, as told by powerful black preachers, of Christian sacrifice. Black children grew up on these stories. Black adults drew strength and comfort from them. And as the Civil Rights movement grew in power, ordinary African Americans, who were filled with the promise of redemption from Christian suffering, found themselves willing to make tremendous sacrifices. Sometimes that sacrifice meant that black families had to sacrifice their own children.

September 15, 1963, was like any other Sunday at Sixteenth Street Baptist Church in Birmingham, Alabama. The only notable event scheduled that day was the monthly Youth Day Service. On those Sundays the church's children had the responsibility of helping people to their seats, handing out church bulletins, and distributing the cardboard fans used to stir a breeze and keep back Birmingham's muggy, heavy heat. Throughout the South black churches stood among the few places where young black people learned to take responsibility as part of a larger cause. The church was where black children got their first chance at public speaking and public singing. In the church, young people often got the opportunity to cut their teeth as leaders and organizers by planning trips or raising money for church dances. If white Birmingham society told black children they were worthless, the black church told them they were indeed special—they were children of God.

Church was so important for these children that, despite the late summer heat, even the youngest arrived in their best Sunday clothes. This was a simple, happy time to be among friends, family, and adults who loved them. They could hardly wait for Sunday school to end so they could take their place in organizing the adult service. Fourteen-year-old Cynthia Wesley was particularly happy that sunny day because it was the first Sunday she got to be an usher. Her friends Carole Robertson and Addie Mae Collins, also fourteen, were serving as ushers that day, too. Denise McNair, their eleven-year-old friend, was following them to share the excitement.

But later one of the Sunday school children told about an eerie feeling as she ran into the sanctuary that morning. Never had she felt so uneasy in church and she concluded that church was too quiet that day. For some reason, she said, a disturbing calm made her feel that something was going to happen to make this day stick out in her memory for years to come.

And she was not alone in her premonition. Just days before a parent with the powerful name of Queen Nunn told her children of a dream so vivid that it left them terrified. She saw death at the door of Sixteenth Street Baptist Church. Bodies were strewn about, and people were screaming and writhing with pain. Bright red blood flowed in the aisles, and the voices of tormented children rang through the sanctuary. This mother later told friends she felt God was trying to send her a signal. At the risk of sounding crazy she pleaded with her children to stay away from Sixteenth Street on that Sunday for fear of "something terrible."

Birmingham, Alabama, had a long history of racist violence. It was a dusty blue-collar town whose steel mills belched clouds of white smoke and whose blast furnaces periodically shook the city with explosions. Throughout the early twentieth century, Birmingham was scarred by labor-related violence as management and unions vied for control of the coal and iron ore mines as well as the steel mills. Often blacks became the focus of violence as they competed with whites for jobs. Some white bosses used the presence of low-cost black workers to threaten the job security of white workers and discourage them from organizing unions. Many white politicians, in the years after the Supreme Court's 1954 decision to integrate public schools, used segregationist speeches—full of bitter visions of blacks taking over schools—to fire up their white supporters.

By the close of World War II, one black neighborhood in Birmingham had earned the name Dynamite Hill because it experienced so many racially motivated bombings. The terrorism was designed to keep blacks away from the city's political and economic power. Among black people throughout the South, the city was called Bombingham because of its racial violence. The city's most prominent civil rights leader, the Reverend Fred Shuttlesworth, had been whipped with chains by white segregationists when he tried to take his children to an all-white public school. Even Nat King Cole, the famed singer, had been pulled offstage by segregationists in Birmingham and badly beaten. In the spring of 1963 Birmingham police had attacked hundreds of unarmed demonstrators, including children, who were simply asking that downtown stores and lunch counters be integrated.

Black churches in the South, as the base for so many civil rights leaders, were the targets of segregationist violence. Between January 1957 and May 1958, there were seven bombings at the homes of black ministers in the South and ten explosions at black churches. In 1958 civil rights leaders from twenty-nine Southern cities met in Jacksonville, Florida, to address the hate bombing issue, especially attacks on churches, and to develop a regional strategy for combating the problem.

In Birmingham the Sixteenth Street Baptist Church was more than a church. It was a focal point for the black community's social and political activity. Sixteenth Street was among the elite of Birmingham's churches. Founded in 1873, it had a proud history of self-help and economic development. Its members had helped establish institutions, from banks and schools to construction companies, that supported the daily needs of Birmingham's blacks. Sixteenth Street's openness earned it the nickname Everybody's Church.

One of the reasons this church was so important in Birmingham's movement was its proximity to the city's central nerve system. The black millionaire A. G. Gaston provided rent-free space in his hotel, which was just down the street. The hotel served as a headquarters for protest leaders. Moreover, the church's location across from Kelly Ingram Park and only a few blocks from the downtown business district made it a logical meeting place for movement participants. The Reverend Wyatt Tee Walker, executive director of Martin Luther King's Southern Christian Leadership Conference and chief strategist of the lunch counter integration demonstrations, meticulously calculated how long it would take groups to walk from the church to various downtown restaurants. His careful planning assured that even if police arrested one set of demonstrators, there was a steady stream of replacements waiting at Sixteenth Street to march and continue tying up traffic and disrupting business.

Even when civil rights leaders met inside the church, hundreds of young people often waited in the park to see what would unfold. The press made no distinction between the people inside and outside the church, so Walker and the SCLC were able to draw numerical strength and momentum from the crowds gathered across the street. By the fall of 1963, the park was infamous as the site where police used nightsticks, attack dogs, and even powerful bursts of water from fire hoses to brutalize black demonstrators. Images of the unholy carnage traveled across the globe of policemen beating young and old people alike. Even white Birmingham businessmen, on a goodwill mission to Japan, reported being shocked by front-page photos in the Tokyo press of

Young protesters defiantly challenge the water cannons
unleashed upon them by the city of Birmingham.

"Magic City" law enforcement officials bullying young protesters in the shadow of the Sixteenth Street Baptist Church.

In August 1963 the Civil Rights movement brought more than a quarter of a million demonstrators to Washington, D.C. Martin Luther King, Jr., in his famous speech on the steps of the Lincoln Memorial, described having a dream of faith in America despite the racial bitterness of the time. He said, "I have a dream that one day down in Alabama—with its vicious racists... one day right there in Alabama, little black boys and black girls will be able to join hands with little white boys and white girls as sisters and brothers. I have a dream today."

Just eighteen days after the March on Washington, King's dream exploded in Birmingham. That Sunday, as Cynthia Wesley and her friends prepared to leave their basement Bible study class, a bomb made of dynamite exploded in the church. Cynthia and her friends

Schoolchildren await transportation to the city jail in Birmingham after their arrest.

Carole and Addie Mae were killed, along with their younger playmate, Denise. The blast left people arriving for church stunned. Some ran holding faces cut by shards of stained glass blown out by the powerful blast. Parents rushed about in panic searching for children who had been in the basement. The parents of those four little girls found that their children had become martyrs to America's civil rights struggle.

Black churches continued to be targets of segregationist attacks throughout the 1960s. A

Participants at the 1963 March on Washington singing "We Shall Overcome," a reworded African American hymn that became the anthem of the movement.

A young girl is consoled by relatives after attending the funeral of children killed at the Sixteenth Street Baptist Church in Birmingham. The bombing at the church was just one example of white terrorism in the South during the Civil Rights movement.

few days before Christmas 1963, a mob of angry white racists burned down the Roanoke Baptist Church in Hot Springs, Arkansas. The church's minister had complained to federal officials about violations of his rights by local segregationists in Hot Springs. His charges got into the local papers. In retaliation the mob set fire to his church. The next year, on June 17, Klansmen beat black worshipers at the Mt. Zion Methodist Church in Philadelphia, Mississippi, then burned down the church. The Ku Klux Klan was worried that the church was the center for a budding Civil Rights movement. But no matter how many churches were bombed or burned, black Christians remained unmoved. They agreed with the prophet Isaiah when he said, "No weapon formed against me shall prosper."

In the summer of 1964 Mississippi experienced a rash of church burnings and bombings. There were twenty-four reported incidents in that state between June and September, and there is little doubt that the spike in attacks was related to the launching of the Mississippi Freedom Summer Project. Organized by the Student Nonviolent Coordinating Committee, the National Council of Churches, and the Council of Federated Organizations, the project brought scores of young people from the North to register black voters and run voter-education drives throughout the state. Black ministers housed project volunteers and permitted the use of their churches for workshops. They were aware that assisting in the movement would make them targets of racist attacks, but many church leaders in the South decided to take whatever risks necessary to further the cause of freedom.

Southern ministers and congregations were not the only ones to answer martyrdom's call. On April 7, 1964, the Reverend Bruce Krunder of Cleveland, Ohio, was crushed when he refused to make way for a bulldozer being used to build a segregated elementary school. His dramatic death drew attention to the fact that in the North whites were willing to build new, segregated schools for blacks. Two years later some white residents of Cleveland burned the newly purchased home of John Compton, a black minister, rather than allowing his family to move into a predominantly white neighborhood. And when a black church opened in a white neighborhood in Providence, Rhode Island, segregationist whites threw Molotov cocktails at the building.

Every attack on a church allied with the Civil Rights movement, be it a black church, a white church, or an integrated congregation in the North, provoked a crisis. The sanctity of the church was being violated, and there were people, even segregationists, who felt that nothing was worth that price. But others argued that violence against churches was a sign that racists feared the potent mix of religious faith and the Civil Rights movement. Black churches stood as targets for the KKK because they represented so much power and credibility among people of all colors. The black church was the one financially independent institution in black American life. Black ministers did not have white bosses. They could not be threatened with the loss of their jobs if they supported civil rights. But as a dangerously charged atmosphere surrounded so many progressive churches and their ministers nationwide, the bombing of the Sixteenth Street Baptist Church was a transforming moment for the movement. In that flash of death, black Birmingham's religious leaders showed that they refused to be intimidated, even when faced with the murder of children.

The most defiant minister was Fred Shuttlesworth of Birmingham's Bethel Baptist Church. Shuttlesworth had long been the strong hand that steered much of the Birmingham movement. Born March 18, 1922, in Mount Meigs, Alabama, a short distance from the state capital, Montgomery, Shuttlesworth grew up dirt poor. But his mother made sure her son had a solid spiritual base and raised him in the African Methodist tradition. When he was three, Shuttlesworth's mother married an out-of-work coal miner, and the family moved to a black mining community on the edge of Birmingham. It was in Birmingham, the largest city in the state, that Fred went to the best black schools Alabama had to offer. And in Birmingham the little boy from rural central Alabama started to develop a political consciousness about the poverty forced on black people who were denied training and jobs. After graduating as the valedictorian of his class, he was determined to do more than go to work in the iron mines.

Shuttlesworth was pulled to the light of ministry. While working at an air force base in Mobile, Alabama, he found himself drawn to the upbeat and "invigorating" music and preaching of the Baptist church. He converted and went to a Baptist seminary. By the summer of 1948 he was preaching at two rural churches outside Selma. His success as an emotional preacher there led him to the pulpit of the most powerful black church in the area, First Baptist Church of Selma.

But Shuttlesworth, a wiry, intense man with a cleft chin, found himself in frequent struggles with First Baptist's board of deacons. The new minister pressed for the church to get involved in politics and the Civil Rights movement. But the church leadership wanted a minister who focused on preaching and left the activism to others. In 1952, while on his way by train to the National Baptist Convention, Shuttlesworth was deeply troubled by the constant tug-of-war with his deacons. "If the Bible is true then all have to suffer and I'm willing," he told God, "but fix me so I won't worry so much." He began to search out a more political church. A year later he resigned as the minister at First Baptist and in March 1953 accepted a call to Bethel Baptist Church in Birmingham.

Within days of his arrival in the big city, Shuttlesworth joined the local chapter of the NAACP as well as the Baptist Ministers Conference. More often than not, he seemed to be a lone voice crying out for the city's churches to join arms with civil rights activists. Encouraged by the 1954 Supreme Court decision ending school desegregation, Shuttlesworth urged Baptist ministers to insist on integrating the local schools. But few seemed ready to answer the call. In frustration, he shifted his energy into the local NAACP chapter's efforts to end school segregation. In 1956 the local chapter of the association elected him membership chairman. His passionate preaching attracted large crowds to their rallies. However, just as Shuttlesworth stirred the local NAACP to action, Alabama's segregationist courts outlawed the group for supposedly inciting violence.

If local laws banned the NAACP, then a new civil rights group was needed. But when Shuttlesworth proposed creating an organization to oppose segregation in the state, he ran into opposition, including a cold shoulder from most black ministers. Several leaders in the Baptist Ministers Conference felt his ambition put the church on a reckless path of confrontation with the state's segregationist business and political leaders. They did all they could to stop him, but Shuttlesworth and a handful of other progressive black clergymen created the Alabama Christian Movement for Human Rights.

Shuttlesworth's action led to sacrifice for him. On Christmas night 1956, he was lying in bed speaking with one of his deacons about future civil rights strategies. During his Christmas sermon he had stated that he was willing to give his life in order to integrate Birmingham. Suddenly a bomb planted under his bed exploded, throwing him across the room and destroying the bed's frame and box spring. The mattress cushioned the blast, and Shuttlesworth was miraculously spared. He was still not intimidated. His personal safety and that of his family had to take a backseat to God's work. He prayed to God and took comfort that his enemies' violence confirmed the importance of his work. He was making a difference in God's name. And he concluded, as did several people around the city, that it had to be the hand of God that had spared him that day.

On the night of the bombing, a policeman on the scene suggested to Shuttlesworth that, for his own safety, he leave town as soon as possible. The activist minister looked the officer in the eye and told him, "You go back and tell your Klan brethren if God could keep me through this then I'm here for the duration." Shuttlesworth's gritty response became instant legend in Birmingham. The man who had worried ministers with his activism was now celebrated as an inspiration.

Four days following the bomb blast that injured his children and was meant to take his life, Shuttlesworth was arrested for sitting in the front of a city bus and refusing to move when asked. And in September 1957 he single-handedly tried to desegregate the Birmingham public schools by enrolling his children at Phillips High School, an all-white school. Before he got to the schoolhouse door, segregationists unmercifully beat him, leaving him bloody in front of his children. His wife was stabbed.

Instead of being humiliated, the bandaged Shuttlesworth preached the next Sunday. With a fierce passion he told his church he was happy that his children had gotten a chance to see him stare down evil, knowing that he was on the side of righteousness and there was no need for him, with God watching over him, to fear any man. He quoted

The Reverend Fred Shuttlesworth regroups protesters at the
Sixteenth Street Baptist Church after they were pummeled by
water cannons and attacked by police dogs.

the passage of scripture that said it is a credit for any man to endure pain while suffering unjustly. If you are able to endure and continue to do what is right, Shuttlesworth shouted to a crowd that spilled out of his church, you have God's approval, and with God on your side you cannot lose. The congregation roared its "Amen."

In 1958 Shuttlesworth and his group turned their attention to the Birmingham police department and its history of harassment of blacks. At the time there were no blacks on the police force, and complaints of mistreatment and outright brutality were constant. It would be nearly a decade before Shuttlesworth could claim a victory in that battle, but widespread anger at the police among blacks in the city allowed him to build an even stronger grassroots organization, with over

600 active members. For the next ten years Monday evening became known as Movement Night, and Shuttlesworth would hold a mass meeting at a local church. White authorities were suspicious of his weekly gatherings, and as the crowds grew, the police decided they needed to be informed about what was being discussed. Police Commissioner Eugene "Bull" Connor and Police Chief Jamie Moore started attending the meetings. Shuttlesworth tried to get a court order to bar them, arguing that their only purpose was to intimidate participants, but his request was denied.

Birmingham's movement continued to push for the desegregation of public schools, lunch counters, and public transportation. In 1963, when Shuttlesworth felt that negotiations with city administrators had reached a stalemate, he invited the Southern Christian Leadership Conference, led by Martin Luther King, Jr., to join the struggle in Birmingham. When Alabama governor George Wallace defied a federal court order to desegregate public schools in the fall of 1963, the black community, again led by Shuttlesworth, responded with protests. At his inauguration earlier that year, Wallace had famously promised a crowd of segregationists that in Alabama it would be "Segregation now! Segregation tomorrow! Segregation forever!" But Shuttlesworth, with only the power of his preaching and appeals to Christian conscience, continued to fight. During a violent clash between police and demonstrators in May 1963, Shuttlesworth was left bloody and with broken bones by police clubs and powerful streams of water. When Bull Connor was told that the preacher had been hurt, he said to reporters, "I'm sorry I missed it."

The demonstrations Shuttlesworth sparked in Birmingham became international news. Pressure grew on the city's businessmen until they agreed to end segregation at stores and restaurants. And when segregationist politicians in Alabama tried to outlaw the new integration,

President John F. Kennedy proposed a civil rights bill to end segregation in all public facilities used by people traveling from state to state, a move that effectively ended segregation when Congress approved it a year later.

While Fred Shuttlesworth reflected the ethic of Christian sacrifice in his work to end segregation in Birmingham, similar sacrifice was taking place in Selma to win voting rights for black Americans. Another man of the church, the Reverend C. T. Vivian, led that struggle.

Born in 1924 in Missouri, Vivian went north with his parents to Illinois. After college he attended American Baptist Theological Seminary in Nashville while working as an editor at the Sunday School Publishing Board of the National Baptist Convention, USA, Inc. He learned about the power of faith-based protest in Nashville by attending Jim Lawson's workshops on Gandhi's nonviolent protests.

By the summer of 1960, Vivian had become one of the guiding forces behind the student-led sit-ins in Nashville. He also was a participant in the Freedom Rides to desegregate interstate travel. As a result of his involvement with that project, Vivian was jailed at the notoriously dangerous Parchman State Prison in Mississippi. An intense man with piercing eyes, the young minister seemed indifferent to intimidation and even the persistent threat of death. Other leaders in the Civil Rights movement said he had an aura of divine purpose. They felt nothing could push him off course, given his history. He was in Birmingham in 1963 when demonstrations led to church bombings. He was in Mississippi in 1964 to participate in Freedom Summer, a project to bring education, health care, and legal rights to poor blacks. It seemed wherever God led him, C. T. Vivian stepped out on faith and responded to the call.

In 1965 God's work led him to Selma. While it was not illegal for blacks to vote in Selma, local election registrars made it difficult for them to sign up. Voter registration offices opened for business only a few days out of the month, and on those days they made sure their hours were limited. The SCLC asked Vivian to coordinate their work in Selma with SNCC and other groups. On February 16, 1965, he led about twenty-five Selma residents to the courthouse to register to vote. It was a cool and rainy morning, and law enforcement authorities would permit only two people inside at a time. Sheriff Jim Clark, dressed in a neatly pressed Eisenhower jacket and military-style helmet, stood guard at the front door with his deputies. His rigid stance made him a perfect target for Vivian and his weapons of nonviolent resistance.

Vivian's most dangerous weapon was a sharp tongue. Clark's physical presence was menacing. But he was clearly outdone by Vivian's verbal skills. Initially, all Vivian asked of Clark was to permit the people to wait their turn in the courthouse lobby so they might get out of the rain. When Clark refused, Vivian drew him into a verbal battle that was caught on tape and viewed around the world. He compared Clark with Adolf Hitler and warned the deputy sheriffs that they would be held morally accountable for their decision to follow Clark. He explained that just as Nazi soldiers who claimed only to have been following orders were eventually hauled into court for crimes against humanity, so would members of Selma's police department be.

The more Vivian talked, the more tense Sheriff Jim Clark became. He felt Vivian was grandstanding for the cameras, and that was true. Civil rights leaders relied on the news media to publicize their causes, as well as to dramatize the evil of segregation and racial hatred. Nonviolent direct action meant that demonstrators like Vivian sought confrontations that would draw out the absurdity of oppression. Nothing captured this drama better than a television camera. While addressing Clark, Vivian occasionally turned toward the members of the media,

The Reverend C. T. Vivian confronts Sheriff Jim Clark
at the Dallas County Courthouse in Selma.

making certain they didn't miss a word. The sheriff accused him of
being an outside agitator, but Vivian asserted that local residents had
invited him to Selma. Moreover, he believed he had an obligation to
confront evil wherever it resided.

Vivian baited Clark into a violent confrontation. He repeatedly
dared the sheriff to show the world how law-abiding citizens were
manhandled in Alabama. In front of TV cameras and photographers,
Clark finally took the bait and punched Vivian square in the mouth.
The minister fell back to the pavement as Clark stood over him, rub-
bing his broken hand. With blood flowing from his mouth, Vivian
never stopped verbally jabbing Clark. "We're willing to be beaten
for democracy," he exclaimed, and "you misuse democracy in the
streets." He pleaded with Clark to arrest him and the others, and
Clark obliged.

Vivian's drama allowed the entire world to witness what blacks in Selma had to endure simply to exercise the right to vote. Vivian wore his fat lip like a badge of honor, happy that God saw fit to use him for the betterment of humanity. The events in Selma pushed the U.S. Congress to pass the Voting Rights Act in 1965. C. T. Vivian's name was not prominent at the signing of the new law, but his spirit and sacrifice were all over its birth.

While ministers like Fred Shuttlesworth and C. T. Vivian walked in the footsteps of Christian soldiers fighting for justice, the black church frequently stumbled on the path to racial equality. Church organizations like the Colored Methodist Episcopal Church and the National Baptist Convention had developed a rather conservative stance on integration. While most within these denominations categorically opposed the humiliating and unequal treatment associated with Jim Crow culture in the South, they differed greatly on the desirable pace and scope of change. Often sacrifices had to be made to get major church groups to join.

The Reverend Joseph H. Jackson, pastor of Olivet Baptist Church in Chicago and president of the National Baptist Convention, was a loud naysayer regarding the Civil Rights movement. Born in Jonestown, Mississippi, on September 11, 1900, he held degrees from Jackson State College and Colgate-Rochester Divinity School. Jackson represented an old guard within the convention that took a conservative posture on the church's role in social change. While he claimed to be a supporter of the movement, Jackson did not advocate the convention's getting involved or taking a stand. He said the church's main business was to win souls for Christ. And Jackson expressed concern that integration might mean the end for all-black organizations—such as the National Baptist Convention. Did Martin Luther King, Jr., and other black Baptist ministers believe that even the separate National Baptist Convention had outlived its purpose?

Jackson had paid his dues and waited his turn to lead, and there was no way he planned to entertain positions that threatened his leadership. He had a reputation of being uncompromising, so anyone who chose to challenge his administration ran the risk of ostracism by those who lacked the courage to stand up to the people in power. A successful challenge would take strong faith in one's position and the belief that the initial unpopularity of controversial stances would not last.

The Reverend L. Venchael Booth, pastor of Zion Baptist Church in Cincinnati, Ohio, knew even before the eventual meltdown in the convention that Jackson's leadership posed a real threat to progressive elements in the denomination. Booth had paid his dues by pastoring small churches in out-of-the-way places.

In 1953, as the Reverend D. V. Jemison prepared to step down as president of the National Baptist Convention, several eager contestants campaigned to succeed him, including emerging leaders of the Civil Rights movement. But the Reverend J. H. Jackson, who opposed church activism, was the victor because very few ministers in the group were prepared to stake the future of the convention on the success of the movement.

By 1957 Martin Luther King, Jr., Ralph Abernathy, and other prominent Baptist civil rights leaders were ready to suggest a new direction for the convention. The Supreme Court had issued its landmark ruling on school desegregation, the Montgomery bus boycott was well under way, and there seemed to be momentum to the movement that had not existed a few years earlier. Booth and his cohorts hoped the National Baptist Convention could be used as a catalyst for the civil rights struggle, but Jackson stood in the way. Their only recourse was the tenure clause in the denomination's constitution. It limited the president of the convention to four consecutive one-year terms.

When the convention met in September 1957 in Louisville, Kentucky, the progressive element, including Booth, pushed the issue of term limits in order to make way for new leadership willing to move in step with the civil rights revolution. Jackson and his supporters, however, had a different agenda. They sought to have the section on term limits lifted from the constitution to permit a fifth term for Jackson. A showdown led to chair throwing and a scuffle between a few opposing clergymen. But when the convention was over, Jackson had succeeded in revoking the term limits clause and been reelected.

By the time the 1960 session of the convention was held in Philadelphia, Jackson had exceeded the constitutional term limit by three years. Booth and other progressive clergymen threw their support behind Gardner Taylor, pastor of Concord Baptist Church of Christ in Brooklyn, New York, and as the election drew closer, the "Taylor Team" gained momentum. When the ballots were counted, Taylor appeared to be the newly elected president. Jackson, however, refused to give up his position. Forces from both sides hurled insults and chairs at one another, and the confrontations became more violent. By the end of the meeting, one minister lay dead, and the final determination about who was the rightful leader of the convention was left to the courts in Philadelphia. But the judge argued that the delegates of the next convention should make the decision.

With court-ordered election procedures in hand, delegates met in Kansas City, Missouri, in 1961 to vote for a president. This time Jackson defeated Taylor and retained his office. He then removed from office a number of elected officials, most notably Martin Luther King, Jr., who had opposed his presidency.

L. Venchael Booth was a loyal member of the National Baptist Convention, but his love for the organization did not overrule his call to God. He felt led to make one final recommendation for the convention's future. He suggested that those who wanted to see the denomination play a more central role in the Civil Rights movement start an alternative Baptist organization. He said the new group would work for the cause of Christ rather than building personal kingdoms.

Again, Booth sacrificed personal friendships and popularity to follow his conscience. He called a meeting for mid-November in Cincinnati for the sole purpose of forming a new convention. But most of Booth's comrades were hesitant to take such a bold stand. The Reverend Herbert H. Eaton, successor to Martin Luther King, Jr., at Dexter Avenue Baptist Church in Montgomery, wrote to Booth, "I shall not hesitate to say how utterly disappointed I am that such an attempt is being made." He believed that further division among black Baptists would do more harm than good to "the cause of freedom." Likewise, the Reverend D. E. King of Zion Baptist Church in Louisville stated, "I do not feel that this is the time nor do we have the spiritual climate for beginning such a movement. For this reason, I cannot participate." And Thomas Kilgore, Jr., pastor of Friendship Baptist Church in New York, wrote, "I feel along with many others that there is much to be done among our Baptist forces, but as much as I feel this way, I am not yet committed to a new Baptist organization."

Although somewhat discouraged by the list of prominent ministers who refused to participate, Booth persevered. In an article titled "A High Call to Greatness," he questioned his comrades' resolve. "Why do men begin crusades without any alternative other than surrender? Are we so poverty stricken in wisdom that we are bound to the weak walls of tradition? Are we over-run by philosophers who cannot bear a rugged cross?" Booth understood the high cost of discipleship, and he stood ready to pay the price. He followed the voice of God as it spoke to him and did not relinquish his commitment to establish an alternative to the National Baptist Convention. His willingness to leave his all

PROGRESSIVE
NATIONAL BAPTIST CONVENTION

The Progressive National Baptist Convention emerged in the 1960s as a response to the controversy surrounding the election of the Reverend Dr. Gardner C. Taylor as president of the National Baptist Convention and the political direction of the convention. First, the incumbent president of the National Baptist Convention refused to relinquish his office to Dr. Taylor even after it was clear that Taylor had won the majority of the votes. Second, the convention was divided over whether it would lend its official support to the Civil Rights movement. If it did support acts of resistance, the convention was faced with determining the level and depth of its involvement in a highly volatile situation. The controversies, which centered on the organization's leadership, triggered a major divide.

In 1961 Dr. L. Venchael Booth, pastor of Zion Baptist Church in Cincinnati, held a meeting at his church with thirty-three delegates and formed the Progressive National Baptist Convention. The organization defined itself as politically progressive and active. Dr. Martin Luther King, Jr., addressed each annual convention until his death. In addition, the organization had its hands in voter registration movements, sit-ins, legislative matters, and economic reform programs.

The Progressive National Baptist Convention has several components, including the Women's Auxiliary, the Home and Foreign Mission Boards, and the Youth Auxiliary. The organization supports educational institutions including the Chicago

Baptist Theological Seminary, Virginia Union School of Theology, and Shaw Divinity School. The Progressive National Baptist Convention organization consists of 1,800 churches and 2.5 million members.

on the altar of sacrifice resulted in the formation of the Progressive National Baptist Convention. Without men and women like L. Venchael Booth, the Civil Rights movement would not have enjoyed the success it did. Leaders often get the credit, but it is the people, those who show resolve in the face of danger, who lead. Without people like C. T. Vivian and Fred Shuttlesworth, the goals of nonviolent direct action would have remained unfulfilled dreams. Their willingness to bear witness to the power of faith in God, to be martyrs for the movement, gave others the strength and determination to persevere.

10

"The Black Messiah"

We issue a call to all black Churches. Put down this white
Jesus who has been tearing you to pieces. Forget your white
God. Remember that we are worshiping a Black Jesus who
was a Black Messiah. Certainly God must be black if he
created us in his own image. You can't build dignity in black
people if they go down on their knees every day, worshiping
a white Jesus and a white God. We are going to communi-
cate with black Churches. We are going to talk to them,
reason with them, shame them if nothing else works, saying,
"Accept the historic fact. Christianity is our religion. The
black Church is the beginning of our Black Nation."

—THE REVEREND ALBERT B. CLEAGE, JR.

Following the passage of the Civil Rights Act of 1964 and the Voting Rights Act of 1965, Martin Luther King, Jr., and other leaders of the black church felt as if God had indeed answered the prayers of the righteous. With the Congress passing major pieces of civil rights legislation and the president signing them into law to be enforced by federal courts, a powerful spirit of optimism flowed in

African American Catholics brought a new dimension of
soulfulness to the ancient church, as is evident at this
revival at St. Sabina Catholic Church in Chicago.

black churches nationwide about the future of race relations. In the mid-1960s the black church stood as the wellspring of much of the civil rights leadership, ranging from King and Ralph Abernathy, to Joseph Lowery and Benjamin Hooks. The black faith community had nurtured these stars who joined the president in the White House for talks on civil rights, urban renewal, and school integration. News reporters solicited their opinions daily, and they were sought after to speak before black and white audiences. The height of this moment came in December 1964, when a child of the black church, Dr. King, was given the Nobel Peace Prize. In accepting the award King said he was acting in Jesus' tradition as a "trustee for the twenty-two million Negroes of the United States of America who are engaged in a creative battle to end the night of racial injustice."

But segregationist opposition to the rising power of the Civil Rights movement remained entrenched. Older white politicians in Southern state capitols and Congress still spoke of defending Dixie's traditions and "states' rights" against what they saw as laws being forced on them by the federal government. And there was doubt in the black community about just how much progress had been achieved. African Americans in the North were especially outspoken in asking how new laws would improve the quality of life in their communities. Cities with large black populations, such as Chicago, Detroit, New York, and Philadelphia, were poised to benefit from increased black voting rights. But it was in the ghettos of those cities that doubts were expressed about the power of new laws to make any significant difference in the lives of working-class black people. Both black and white politicians and religious leaders feared that America's urban centers were time bombs, waiting to erupt in riots at the slightest provocation.

The sharpest voice doubting advances made by King and the movement came from the Nation of Islam. Often taking aim directly

Elijah Muhammad and Martin Luther King, Jr.,
meeting during the Civil Rights movement.

at Dr. King and the Southern Christian Leadership Conference, Elijah Muhammad continued to point to what he called the futility of trying to love white people into submission. Muhammad believed that King had been tricked by the white man into believing that whites and blacks have a bond of brotherhood through the Creator. In his 1965 book, *Message to the Black Man in America,* Muhammad made it clear that blacks would never be God's "select people" as long as they embraced the white man's religion, Christianity, as well as the politics of an integrationist agenda. "The American so-called Negroes are gravely deceived by their slave-masters' teaching of God and the true religion of God," he said. "They do not know that they are deceived and do earnestly believe that they are taught right regardless of how evil the white race may be.... You are made to believe that you worship the true God, but you do not!" No amount of education, federal legislation, or political agitation could change the condition of blacks in America, he concluded. Only a proper relationship with the black man's true God—a black God—could fix all the problems that ailed blacks and put them on the throne of leadership.

Muhammad's problem was that he had few followers. Even black people who sympathized with his defiance were reluctant to give up the religion of their forefathers, a religion that had sustained black people through centuries of persecution and oppression. But as the Black Power movement emerged in the mid-1960s, the message of Elijah Muhammad began to make new inroads into black communities in big cities of the North and especially among the poor, the drug addicts, and the black people in America's jails. The Black Muslims offered hard, clear answers about self-reliance and self-respect to black people who felt the new civil rights laws made no difference to them. One prisoner who was touched by Elijah Muhammad's message was a thin, reddish brown man named Malcolm Little.

Known to people on the street as Detroit Red, Malcolm was a small-time hustler who ran afoul of the law while living in Boston. Born May 19, 1925, in Omaha, Nebraska, he was the son of J. Earl Little, a construction worker and self-ordained Baptist minister, and M. Louise Norton, West Indian immigrant from the island of Grenada. Malcolm's father was a staunch black nationalist preacher and committed follower of Marcus Garvey's Universal Negro Improvement Association. Life for young Malcolm was extremely rough. When he was six years old, his father was found dead in the streets of Lansing, Michigan, apparently hit by a trolley. His mother was left to raise him and his brothers and sisters by herself. Later his mother told Malcolm that whites killed his father. The loss of her husband coupled with the pressures of holding her family together proved too much for her, and eventually she wound up in a mental hospital. Malcolm and his siblings were shuttled in and out of foster care, where they internalized deep feelings of shame and worthlessness. By the time he had moved to Boston to live with his older sister and her husband, young Malcolm had already decided to live on the streets. The world had given him nothing but a raw deal, so he felt no need to look out for anyone but himself. His anger and pain were slowly consuming him, and he sought refuge not in God but in drugs, alcohol, and women.

Malcolm's life of crime as Detroit Red spiraled out of control. He was convicted and sent to prison. He wasn't even twenty-one years old, but he was filled with cynicism. Fellow inmates recognized in Malcolm the self-loathing that masqueraded as fearlessness and bravado. His strong antireligious position earned him the nickname "Satan" among the prisoners. But through the encouragement of other men in prison, he began to educate himself. He started by reading the dictionary, then he devoured every piece of literature he could get his hands on. If he was ever going to be a changed man, he knew he would need knowledge of more than picking locks and other petty crimes.

One day Malcolm's brother Philbert wrote and told him about the teachings of Elijah Muhammad. Malcolm was unimpressed by what he read but pleased that his brother was trying to straighten out his life. Soon after receiving Philbert's letter, Malcolm was visited by another brother, Reginald, who also spoke with a revivalist fervor about the good being done by the Black Muslims. He told Malcolm that "God is a man" whose "real name is Allah." Malcolm was confused but intrigued. Was the Nation of Islam another street hustle or true salvation for his family and other black people? To find out more about the Black Muslims and their leader, the man known as the Messenger,

Malcolm asked his other siblings (Wilfred, Hilda, and Philbert), who also had converted to the Nation of Islam, to bring him information on the group. They taught him about the beliefs of the Nation of Islam and the many ways it had changed their lives. The more Malcolm learned about the Nation of Islam's views of the world and a black God, the more he wanted to know. He became convinced that the key to freedom from the psychological and spiritual prisons that kept so many black Americans poor and ignorant was to be found in the Nation.

When his family told Malcolm that all he needed to do to get paroled was follow the rules of the Nation of Islam, he saw it as a challenge. He began to change his diet, avoid drugs, and clean up his life. But the transition from hustler to holy man was more difficult than he had anticipated. One of his first stumbling blocks was his inability to pray. Although his father had been a Christian preacher, Malcolm was bitter about God. Since God had never seemed to be there for him, there had been no need to keep open lines of communication between himself and the Creator. He wasn't so sure God even heard prayers, let alone answered them. But the Nation of Islam required that he submit himself to the will of Allah, and prayer was the first act of emptying himself and allowing Allah to lead his life.

In his autobiography Malcolm described that process. He understood the mechanics of prayer, but something prevented him from following through. He wrote, "The hardest test I ever faced in my life was praying.... My comprehending, my believing the teachings of Mr. Muhammad had only required my mind's saying to me, 'That's right!' or 'I never thought of that.'" Actually getting on his knees, however, took him the better part of a week. Malcolm's life of crime was now an embarrassment to him, and the only way to be released from the burdens of the past was to confess his transgressions to Allah. "For evil

A young Malcolm X, national spokesman for the Nation of Islam.

to bend its knees, admitting its guilt, to implore the forgiveness of God, is the hardest thing in the world," said Malcolm. "It is easy for me to see and say that now. But then, when I was the personification of evil, I was going through it." Once he was able to tell Allah about his troubles, he was freed from the weight he had carried. From that point forward he committed himself to a life of righteousness under Elijah Muhammad. He would spend the next decade teaching others about Elijah Muhammad and what he called the true religion of the black man.

Once paroled, in 1952, Malcolm became a Nation of Islam minister. He changed his last name from Little to the letter *X*, to symbolize his missing cultural heritage. He said his true identity had been stolen by the white devil and could never be adequately replaced. He dismissed Little as a slave name given to his forebears by their white owners. By shedding his slave name, Malcolm X said, he also shed his slave mentality. By 1954 Malcolm had become a commanding and witty speaker. But he was more than an orator. His reputation as an organizer, recruiter, and leader grew rapidly, drawing attention from the press, politicians, and even the police, who worried about the presence of Malcolm's disciplined men in Harlem. Elijah Muhammad was so impressed with Malcolm that he named him minister of Temple 7 in Harlem. This was familiar territory for Malcolm. He had spent many days hanging out on Harlem's streets and in bars looking for trouble. But that seemed like another life to him; now Malcolm was living proof of the transformative power of a committed faith. Not faith in some "spook God," he said, who was content to be shrouded in mystery, but faith in a God who loved his own people enough that he revealed his true identity to them, walked among them, taught them, released them from the shackles of ignorance, and exposed them to the true power of being black. Malcolm had a God that chose sides, and that God clearly stood with blacks.

In the 1950s and '60s Harlem was pulsing with urban black life. There seemed to be a preacher or prophet fishing for members on every street corner, and Malcolm was not hesitant about casting his line into the waters. Although Harlem was known all over the world as the cultural capital of black America, where the best of jazz could be heard in nightclubs and the finest gospel burst from churches, Malcolm was in touch with the gritty underbelly to Harlem's image. He knew the back alleys and dope dens, and he was the one minister who had no fear of walking into the darkest joint to spread the word of the Honorable Elijah Muhammad. If it could save him from a wasted life, he contended, it could surely work for his listeners. But Malcolm X's message was a tough sell in the face of easy thrills, such as liquor, drugs, and gambling.

There was also open resistance from Harlem's black aristocracy, including the mainline Baptist and AME church leaders. New York City was also a place of social and economic segregation, and Harlem's churches reflected that divide. Streets like West 139th Street, known as Strivers Row, offered havens for middle-class blacks seeking the social distinctions of bourgeois culture. Malcolm knew that Harlem was not the cohesive community it pretended to be, and he worked to find his members among those who felt rejected by established churches. He attacked local black political and civil rights leaders for not doing more for black people struggling with poor jobs, brutal police, and corrupt politicians. Black churches came in for criticism, too. He said they were brainwashing black people to worship a white God. He sought out unemployed black people just up from the South, for whom Harlem was not the Promised Land but rather hell on earth. Instead of the dope he used to peddle on 125th Street, Malcolm was now selling religious faith.

As Temple 7 grew, so did Malcolm's status as a black leader in New York. Then he got a surprise promotion. Elijah Muhammad, in Chicago, named him national spokesman for the Nation. A new challenge came with the new title. Jealousy erupted among the ministers who thought he was getting too much glory. Malcolm, however, was not concerned, for he never failed to give the praise and honor to Allah and his messenger, Elijah Muhammad. With his faith firmly rooted in the conviction that it was his duty to save as many blacks as he could from the lies of the white man, he remained steady in his work.

As the Civil Rights movement continued nationwide, Malcolm X and the Nation of Islam stood apart from the sit-ins, marches, and other nonviolent protests held by student groups as well as by Martin Luther King, Jr., and the Southern Christian Leadership Conference. Although not an advocate of violence, Malcolm had no use for King's preaching about the black man turning the other cheek. What was the value in loving one's enemy? How could anyone in his right mind love someone who hated him, who had lynched his neighbors and burned crosses to terrorize black families? King's philosophy was rooted in the notion that whites had a conscience that could be reached, transformed, and redeemed. Malcolm did not buy that logic, for the Nation of Islam taught that the white man was the Devil, and it is counterintuitive for the Devil to be anything but devilish. Elijah Muhammad said the white man had no soul and no life worth redeeming.

Malcolm X believed black civil rights leaders would be better off trying to teach black people to love themselves and not their enemies. This message resonated with young blacks and white college students who were unfulfilled by marches and demonstrations. Even though Malcolm X's followers were only a sliver compared with King's national following, major American newspapers and the television networks gave Malcolm X regular coverage as the bitter, violent opposite of the peaceful Dr. King.

While Malcolm X and the Nation of Islam tried to demonize all white people, they could not adequately account for white students, ministers, and political leaders who locked arms with blacks and risked their personal safety. White participation in the Freedom Rides and the Selma march in the summer of 1964 showed blacks and the world

The only meeting between Dr. Martin Luther King, Jr., and Malcolm X.

that not all white Christians were rabid segregationists. A significant number were willing to stand with blacks in their struggle for equality and justice, and this was an encouraging sign that America might live up to its promise to be a nation of "liberty and justice for all." Malcolm X told white college students the best thing they could do to help blacks was work on racism in their own houses and neighborhoods and let the black man stand on his own feet.

Ironically, black people killed Malcolm X while he was speaking in Harlem in 1965. A struggle between Malcolm X and Elijah Muhammad provoked the fatal gunfire. Malcolm had gone to Mecca to experience authentic Islam and returned with new visions of unity across racial lines. He questioned Elijah Muhammad's teachings about the white man as a devil and Muhammad's extramarital relationships with women in the Nation. The feud led to Malcolm's assassination by members who viewed his post-Mecca transformation as a betrayal.

Malcolm's murder and later riots created disillusionment in Northern cities. When Dr. King was assassinated three years later, black America, particularly young activists, began to question whether the love ethic would ever take root in a land that seemed to understand only violence. If the love ethic was ineffective, then King's philosophy of brotherly love and nonviolence had lost its power as a strategy for achieving social change. As doubts took hold, the Civil Rights movement began to push past the black church and even the extremes of the Nation of Islam. More and more young people wanted their freedom now, not later. They wanted economic redress for centuries of unequal treatment now, not later. They wanted better schools for their children now, not later. And if being a Christian meant having to wait on the Lord to solve their problems, then more blacks than ever were prepared to say "to hell with that religion."

A small but vocal number of young black Christians experienced a profound crisis in faith. Was Christianity more receptive to the needs and desires of whites? How could blacks be Christians, given all the treachery committed against black believers by white Christians? Was there something inherently wrong with the faith itself, or was the problem its misuse? Could Christianity be reconciled with the Black Power movement?

While most of mainstream Christian America, black and white, found such questions unwarranted, a handful of black clergy sought a meaningful response to critiques from the Black Power movement and the Nation of Islam. On July 31, 1966, the National Committee of Negro Churchmen issued a statement on black power that tried to reconcile the movement with the mission of Christianity. Aiming their comments at white Christians who chose to dismiss demands for black power because they linked it with hate and violence, the committee stated, "We regard as sheer hypocrisy or as a blind and dangerous illusion the view that [love and power are opposites]. Love should be a controlling element in power, not power itself. So long as white churchmen continue to moralize and misinterpret Christian love, so long will justice continue to be subverted in this land." While the committee was certainly progressive on the issue of black power, their views represented a small portion of black Christians. Black power was still a relatively new idea, and most pastors and parishioners wanted no part of the aggressive tone of a black power movement. They saw it as incompatible with the teachings of Jesus, although some tried to make room for the new ideas that were transforming the Civil Rights movement.

No minister was more vocal about the need for black Christians to advocate black power than Albert Cleage, Jr. Born in Indianapolis, Indiana, on June 13, 1911, Cleage grew up in Detroit, where his father was one of the city's pioneer black physicians. His light skin color and class made him a part of Detroit's colored aristocracy, but young Albert Cleage seemed consistently to gravitate toward issues affecting the largely unskilled masses. His family had a strong commitment to education, so there was never a question of whether he would pursue a college degree. He went to Wayne State and Fisk Universities, as well as the Oberlin Graduate School of Theology.

As a young man, Cleage was inspired by the labor and civil rights struggles in his hometown, and throughout his professional life he held radical views on religion and politics. During the 1940s Cleage preached a doctrine of integration, and in 1944 he accepted a temporary position as copastor of the newly formed Fellowship Church of All Peoples. An experiment in integrated worship, the Fellowship Church hoped to prove that people of all cultures could come together under the inspiration of Christianity and work for the common good of American society. Its founders, including the black theologian and preacher Howard Thurman, wanted to show the world that Christianity was not a racist religion, but that, given the right conditions, it could thrive as an authentically interracial enterprise.

Cleage, too, was committed to interracial fellowship, but not at the expense of either group's rights. When he arrived in California, the United States was still at war. A large portion of the Fellowship Church were Japanese Americans, all of whom had been forcibly removed from their homes and were temporarily housed in internment camps. The racial tensions in San Francisco were as high as in other American urban areas. And while it was able to escape the racial violence that afflicted Detroit in the summers of 1942 and '43, San Francisco was but a few degrees away from its boiling point. The interracial Fellowship Church was well intentioned, but Cleage quickly came to feel it was a "contrived, artificial affair" that did not address the unequal distribution of power between minorities and whites.

The Reverend Albert B. Cleage (1911–2000) founded the Shrine of the Black Madonna, a nationalist Christian movement. In two books, *The Black Messiah* and *Black Christian Nationalism,* Cleage chronicled his attempt to formulate a black theology for African American Christian churches.

Although he was initially in agreement with the values and principles of the church, he became increasingly uncomfortable with the mix of people. He grew to feel that true integration was impossible. How, he asked, could racial unity be discussed with any honesty without first figuring out what to do about property that had been taken from the congregation's Japanese American families? Basic justice issues had to be dealt with before this Christian experiment could have any relevance in the lives of parishioners. His questions about the political landscape of San Francisco earned him a reputation as an extremist among the church leadership. Even Thurman believed that the mission of the Fellowship Church should be creating common ground through interracial spirituality. In 1945 Cleage left San Francisco disappointed by the unwillingness of white Christians to move beyond mere advocacy of desegregation and toward social integration.

For the next six years, Cleage was pastor of St. John's Congregational Church in Springfield, Massachusetts. This was a solidly middle-class congregation with a history of social involvement dating back to nineteenth-century abolitionism. Cleage initially felt right at home. But Springfield was no Detroit, and even though he had a productive ministry at St. John's, Cleage searched for an opportunity to get back home. In May 1951 he accepted a call to serve as pastor of St. Mark's United Presbyterian Mission, a black middle-class congregation in his hometown. Although he was ordained a Congregational minister, Cleage was no stranger to Presbyterianism. And his father and uncle were charter members of St. Mark's, so he had little reason to believe his tenure there would be anything other than smooth.

Nonetheless, he soon ran into conflict with the local synod over his governance of the church. Unlike the Congregational churches he'd pastored, Presbyterian churches had a rather rigid hierarchy that involved significant oversight by local denominational officials. Cleage's tension with the synod exemplified some of the pitfalls that hindered the growth of blacks in predominantly white denominations. While he was certainly happy to accept denominational resources, Cleage resented outside interference with his parish's activities. His days at the Fellowship Church had taught him that religious cooperation and integration usually had to happen on whites' terms, and he was not willing to sacrifice the needs of his flock. Why should some white man from the suburbs have any influence over a congregation in inner-city Detroit?

Once again Cleage seemed to be ahead of his generation, for it would be a full decade before blacks in the Presbyterian and Congregational churches made the requests of their denominations that Cleage made in the 1950s. Demanding freedom to run the congregation according to the needs of the local black community and not the church's national structure, he led "a group of dissidents" out of St. Mark's in March 1953 and formed the Central Congregational Church.

Cleage spent most of the next decade trying to establish a firm foundation for his newly formed church. Between 1953 and 1958 the group called several venues, including a large house and an old school building, home. Even though it was difficult starting a church virtually from scratch, Cleage enjoyed the relative independence that came with the Congregational style of church order. Before long he had built Central into a small but strong church whose work went far beyond the building's four walls. Cleage believed that following Jesus required a commitment to the needs of the downtrodden, and a middle-class congregation like Central could fulfill its mission only by sharing its blessings with the less fortunate. From soup kitchen to social programming and job training, Cleage believed it was the job of the church, particularly local congregations, to take the lead on issues of community development. Once Central was on solid ground, Cleage was able to devote more of his time to various community organizations. He was particularly involved with the National Association for the Advancement of Colored People, as a member of its executive board. It was not long, however, before Cleage began to pull away from groups like the NAACP and affiliate with groups more critical of not only the white establishment but the Negro bourgeois leadership as well.

Early in 1962 Cleage played an integral role in organizing the Group of Advanced Leadership. GOAL, as it was called, was made up primarily of young black leftists who advocated new directions for the Civil Rights movement in Detroit. Thus, it was an honor for Cleage, at the time fifty years old, to be elected GOAL's director. A little over a year later he was part of the leadership of another group aimed at reshaping the black agenda in Detroit. The Detroit Council for Human Rights, formed on the twentieth anniversary of the Detroit Riot of 1943, posed the most serious threat to the established Negro leadership, partly because of its credibility in the eyes of both black and white communities. The Reverend C. L. Franklin, pastor of New Bethel Baptist Church, was the group's president.

Because of DCHR's somewhat prominent membership, it was able to accomplish things that groups such as GOAL could never achieve. On June 23, 1963, after only one month of existence, DCHR pulled off the Walk to Freedom, up to that point the largest civil rights march America had seen. Cleage was one of the march's directors, and the Reverend Dr. Martin Luther King, Jr., came to Detroit to participate. Even though DCHR had been very successful in a short time, Cleage never gave up his work with far left groups like GOAL. On October 11, 1963, GOAL organized a Michigan chapter of the Freedom Now Party, a national political organization founded in Washington, D.C., in August 1963 after the March on Washington. Freedom Now's primary objective was to get blacks elected to political office, particularly on the local level. The hope was that blacks could gradually gain control of the institutions that most affected their lives. While Cleage was, for the most part, both willing and able to balance his commitments to DCHR and GOAL, the time soon came when he was forced to choose sides.

Although he did not discount DCHR's ability to effect change within the existing social parameters, Cleage believed his calling as a minister of the gospel compelled him to attack injustice at its roots. Just as he had in San Francisco, he advocated radical approaches to social change. For example, he believed the Southern Christian Leadership Conference under the direction of King was ill equipped as a nonviolent organization to respond to the challenges of racial inequality in Northern cities.

The test of Cleage's commitment to black religion and black radicalism came in November 1963, when C. L. Franklin convened a meeting in Detroit of Northern civil rights activists. The Reverend Adam Clayton Powell, Jr., a U.S. congressman from Harlem and pastor of the renowned Abyssinian Baptist Church in New York City, was to be the keynote speaker. Cleage was excited about Powell's participation. Powell's commitment to race over party was unquestionable, and Cleage believed leaders like Powell had the power to open the national dialogue to a wide range of black political perspectives. However, Cleage soon realized that Franklin "would not permit the conference to give any consideration to the all-black, pro-separation Freedom Now party." If there was to be a Northern Christian Leadership Conference, it clearly would not include the radical views of Christians like Cleage. In response, Cleage resigned from DCHR and accepted the chairmanship of Michigan's Freedom Now chapter.

Cleage had taken a stand in claiming that black power indeed was compatible with Christianity, but not everyone appreciated his position. Some of his parishioners at Central Congregational were unhappy with his fusion of religion and leftist politics. They supported his efforts to make the church more relevant in the lives of Detroit's needy residents, but they believed his support of a separatist agenda in the Motor City went too far. Following an October 1963 Freedom Now rally held at the church, the *Michigan Chronicle,* Detroit's major black newspaper, reported that several unnamed church members hoped to have Cleage dismissed from the pastorate. However, since no one confronted Cleage personally, he felt he had no need to worry.

OPERATION BREADBASKET

In 1961 the Southern Christian Leadership Conference established an economic component called Operation Breadbasket. The economic division of the civil rights organization, founded by Dr. Martin Luther King, Jr., was designed to provide economic stability and wealth to African Americans. It encouraged and trained ministers to organize boycotts against businesses perpetuating racism and discriminatory employment practices against blacks.

Though Operation Breadbasket was housed in many cities, its biggest success occurred in Chicago, when the then student leader Jesse Jackson worked with the Reverend Clay Evans and other local ministers to help found the local chapter in 1966. Jackson and others led successful campaigns against several leading retailers, including A & P grocery stores. Jackson's polished leadership style garnered the attention of the SCLC's founder, and Jackson was appointed national director of Operation Breadbasket.

A number of changes in American politics and shifts within the movement caused the steady decline of Operation Breadbasket. Jackson resigned in 1971 and founded People United to Save Humanity (PUSH).

Unfortunately, the matter went well beyond the black newspaper and into the ears of white trustees of Detroit's Metropolitan Association of the United Church of Christ. At the request of two Central Congregational members, the association held a hearing on June 8, 1964, to determine whether Cleage's teachings on black power were within the bounds of Christianity. If the ruling went against Cleage, the association was prepared to consider withdrawing financial support until the parish dealt with its pastoral problem.

A memo written by his anonymous critics was read into the minutes, and Cleage was asked to respond. His detractors outlined several key issues they wanted Cleage and the association to resolve:

> He takes, we believe, at least six positions which should be considered seriously and urgently by all who call themselves Christian: (1) nonviolence has failed; (2) the federal government is anti-Negro; (3) racial conflict is inescapable; (4) integration is not the goal; (5) black nationalism, not Christianity, is the vehicle for achievement; (6) black-white cooperation is rejected.

Before Cleage could attest to whether the information in the memos was true, Thomas Williams, a member of Central and a spokesman representing the thirty or so Cleage supporters at the meeting, requested that a letter from Cleage loyalists be included in the minutes. The letter stated that it was not the place of the association to serve as judge and jury regarding the internal matters of an autonomous institution. There was no doubt that Cleage was teaching black nationalism—"no question about it." However, they believed Cleage's black nationalism was Christian.

To Cleage the probing of his ideology exemplified the worst of white paternalism. As for the complaint that he preached hatred for whites, he stood to respond: "White men bore me, I don't hate them. I never preach hate. They bore me to tears. . . . You hate me, though I love you. You wish I hated you; it'd be easier for you. I feel sorry for you." The tension in the room was palpable, and Cleage did little to calm the situation. He wanted those who stood in judgment of him to know that he was not afraid to speak his mind in the presence of whites, and he refused to back down from his general claims about the necessity for blacks to have a Christianity that represented their life experiences and struggles.

After the buzz in the room quieted, he answered what some might have seen as the two most serious "charges" leveled against him: that he publicly said Martin Luther King, Jr., was a failure; and that he appeared on the same platform with Malcolm X and other Negro radicals. To the former complaint, Cleage said it was no secret that King's tactics had failed, citing what he believed was a lack of "enduring change in society" as a result of Southern civil rights campaigns. Sure, black folk in Nashville and Birmingham could eat at formerly whites-only lunch counters, and anybody could sit in the front of the bus in Montgomery, but blacks and whites were just as far apart as they were the day Rosa Parks was arrested. The best-paying jobs were still reserved for white workers, many communities in the Detroit metropolitan area still did not welcome black residents, and it appeared that all the marching in the world would not change the harsh reality of racial separation in America. To the latter charge Cleage replied, "Yes, [I appeared with Malcolm] and you'll note in reports that I gave a Christian invocation and I did it as a Christian clergyman, and you ought to have Malcolm out to your churches...he could do you so much good." Cleage was determined not to apologize for his politics or for his belief that Christianity and black nationalism were compatible.

When the smoke cleared, Cleage was still the pastor of Central Congregational Church. The church and ministry committee voted four to one not to request further investigation of his ministry. However, this outcome did not resolve the discontentment in the church. After all, some person or group had gone behind Cleage's back to the Metropolitan Association. It was not long before his accusers were made known. Thirty people withdrew their membership from the church.

Firm in his belief that black Christian nationalism was an appropriate and necessary response to all that ailed black America, Cleage refused to be dissuaded by the ostracism of mainstream religious and political organizations. He was convinced that it was a "new day in Babylon," which required new approaches to black spiritual strivings. At the height of the influence of the Civil Rights movement, Cleage was telling all who would listen that King and nonviolent direct action were unable to alter the course of evil among white racists, particularly in the North. While integration and racial cooperation were still important buzzwords among liberal Christians, Cleage was teaching that if whites would only allow blacks to control the institutions that affected them, integration would be neither desirable nor necessary. Cleage's theology and ideology continued to evolve. By the end of 1964 it was clear that black nationalism was at the heart of his Christian message.

And that message of black religious radicalism swung further to the left as the Black Power movement gained momentum nationally. In the summer of 1966, around the time Stokely Carmichael was issuing calls for black power in Mississippi, Cleage held the First Annual Black Arts Conference at Central Congregational. It was his hope that the conference would draw attention to the need to accept the Nation of Islam's challenge for blacks to find beauty in their own identity and culture. Cleage did not limit the need for an aesthetic revolution to secular institutions. Indeed, he believed that black churches were the biggest perpetrators of racial self-hatred in the black community. With images of a white Jesus on the walls of nearly every black congregation, Cleage did not see how blacks would ever be free.

The following year's conference attracted black nationalist figures like Carmichael and H. Rap Brown. Cleage's articulation of a black aesthetic culminated in the unveiling of a painting of a Black Madonna on Easter Sunday, March 26, 1967, and the renaming of Central Congregational the Shrine of the Black Madonna. The eighteen-foot painting of a dark-skinned, barefoot, black Mother of Jesus, cloaked in a flowing blue robe with a white veil covering her head, and holding a jet black baby boy in her arms, bore no resemblance to popular European American representations of Mary. Her furrowed brow, intense gaze, and muscular forearms reflected a woman who had known hard times and could hold her own in a crisis. She was prepared to do whatever was necessary to protect her child and her people.

There was no mistaking Cleage's point—Jesus was a black man, a revolutionary sent by God the Father to liberate the oppressed black nation of Israel. Cleage wanted to "dehonkify" Jesus and Christianity so blacks could feel it was their religion, not one given to them by whites for the purpose of holding blacks down. On July 23, Detroit erupted in violence after police raided an after-hours nightclub in the black community. By the time the governor called in National Guard units to quell the rebellion, 43 people were dead, 1,189 injured, and 7,231 had been arrested. Detroit would never be the same. With its infrastructure scarred and the threat of future outbursts looming, white residents fled Detroit for the suburbs. An economic slowdown and a rise in joblessness were among the many causes of the unrest, and white residential and economic flight only exacerbated the situation.

The black proportion of Detroit's population increased from 29 percent in 1960 to 45 percent by 1970.

As Detroit grew blacker and poorer, Cleage felt that the need to make Christianity relevant to the masses of urban blacks was stronger than ever. They needed a Jesus who looked like them, had experienced racial discrimination at the hands of Europeans as they had, and was ready to fight along with them for the empowerment of their communities. He spent most of 1968 writing and promoting his book *The Black Messiah*, which includes several sermons he delivered at the Shrine of the Black Madonna, along

Wayside Madonna, by Edith C. Phelps.

with a few articles reprinted from his weekly column "Message to the Black Nation," which ran in the *Michigan Chronicle*. In addition to his book, his views became public through his work on the board of the Inter-Religious Foundation for Community Organization and as an adviser to the Detroit-based National Black Economic Development Conference, which was run by the former Student Nonviolent Coordinating Committee leader James Forman.

In 1972, with the publication of his second book, *Black Christian Nationalism*, Cleage inaugurated the Black Christian Nationalist Movement as an independent denomination. The first official name for the denomination was the Black Christian Nationalist Church, Inc.; it was later changed to the Pan-African Orthodox Church. By 1972 Cleage had also taken the name Jaramogi Adebe Agyeman; Jaramogi means "liberator, holy man, savior of the nation" in Swahili. He is addressed as the Holy Patriarch of the church, which currently has three congregations in the United States. Upon learning about Cleage's new church, the Reverend James H. Hargett, secretary for Black Ministries of the UCC Council for Church and Ministry, stated:

> Only if one had read Albert B. Cleage's book, *The Black Messiah*, is it possible to understand why he had to conclude that if form follows function his practical theological position necessitated a new form for fulfillment.... I hope Jaramogi Adebe Agyeman of the Black Christian Nationalist Church, Inc. will retain his standing in the United Church of Christ. He has contributed brilliantly to the UCC Black Caucus' demands for increased opportunity for black members within the United Church of Christ, and has helped make the church more sensitive to and aware of its need to respond to the agenda of black people.

Cleage's congregations have retained affiliation with the UCC and continue to serve as sources of inspiration and pride for many people in the black community.

While black Christian nationalism never really caught on the way Cleage hoped, it set the stage for the emergence of a more Afrocentric Christianity in the 1980s and '90s. The experiences of the 1960s and the Black Power movement taught black religious leaders that the old-time religion might not be good enough for the young people of the late twentieth century. If Christian churches were going to hold the attention of their youth, they would have to adapt to changes in American popular culture. Churches that chose not to make changes have not experienced the same rate of growth as those that have. Some black leaders came to reconsider Christianity. Throughout the 1980s and '90s, young religious leaders sampled various faith traditions to come up with an eclectic spirituality that was less formal but more practical. Their faith was expressed not through creeds and doctrines but through the clothes they wore, the music they listened to, and the food they ate. The faith of the hip-hop generation is grounded in the reality of the streets.

11

Fishin' for Religion

The Middle Passage—chained enslaved Africans in the
holds of several ships of every Atlantic maritime nation—
was never forgotten by the Africans, neither during slavery
nor in freedom. The watery passage of the Atlantic, that
fearsome journey, that cataclysm of modernity, has served
as a mnemonic structure, evoking a memory that forms the
disjunctive and involuntary presence of these Africans in the
Atlantic world. From this perspective, religion is not a cul-
tural system, much less rituals or performance, nor a
theological language, but an orientation, a basic turning of
the soul toward another defining reality.

—CHARLES H. LONG

In 1970 nearly 34 percent of all black Americans lived below the poverty line. The American economy was in a recession, and the black urban neighborhoods that had once attracted migrants from the rural South now resembled war zones filled with violence, crime, and despair. Riots in several cities left many black leaders, including church leaders, wondering if the Civil Rights movement would ever

Black youths holding their fists in the air
as a symbol of the Black Power movement.

recover from the seemingly uncontrollable rage of black power. The young men and women who made up the Black Power movement saw things differently. They saw the outcry from their brothers and sisters as a response to the slow pace of change in America. To them, their future looked bleak. Manufacturing industries were trying to increase profits by cutting labor forces, which resulted in fewer high-wage jobs, and employment in the service sector did not pay enough to make a decent living. Moreover, young black men were drafted to serve in the armed forces at a rate disproportionate to that for whites, with only 30 percent of eligible whites inducted compared with 60 percent of eligible blacks. For the young black men who were pressed into military service, America's war in Vietnam was "a rich man's war but a poor man's fight."

Despite the overwhelming anxiety, faith prevailed in black communities throughout America. But, unlike in previous generations, Christianity shared the stage with many other religious traditions. While younger black folk continued to believe, they were more willing than ever to explore new dimensions of faith. Princeton theologian Dr. Cornel West wrote:

> You can have faith without being a religious faith. Faith is the notion that somehow you can muster the courage to step out into the unknown and still sustain yourself, or be sustained. Christian faith is stepping out on nothing, landing on something, because that something is associated with a higher power than you...Jesus in the New. So that when we think of faith, there's so many different forms in which faith is manifest.

At the end of the 1960s, faith among black people was as vibrant as ever. But questions about the relevance of Christianity to the black experience in America were being voiced more loudly than in the past.

The Vietnam War added to the rush of questions about Christian life. Black soldiers returned home with Buddhist prayer beads on their wrists. The American peace movement also brought black people in touch with Eastern faith traditions and expanded the range of possibilities for those in search of new beginnings.

Hinduism and Buddhism became new alternatives on the black religious landscape. The fact that these faith traditions did not carry the history of racism against blacks made them all the more appealing to those seeking a way out of Christianity's tangled racial and economic caste system. This did not mean there were no racists practicing these religions, but it did mean that the teachings and sacred texts themselves did not imply that blackness was a curse. Black American attraction to Buddhism and Hinduism remained rather small, however, for the majority of black seekers were unprepared to give up their cultural identity in order to find God. They had just discovered the power of what it meant to be black, and surrendering that power for a deracialized religion seemed naïve to them. But, as author Angel Kyodo Williams wrote:

> While we don't necessarily have to be looking to replace our traditional religions, we can appreciate the benefit of weaving some of the very useful practices and ideals of ancient Eastern wisdom into our daily lives. Just as we have recollected the valuable aspects of the African-derived belief systems of Yoruba, Santeria, and even Voodoo, many of us have also taken the liberty of exploring Islam, Hinduism, Sufism, or one of the different schools of Buddhism.

Some blacks have found the change to Eastern religions both useful and necessary.

One person who was willing to try something different was Anna Mae Bullock, known to the rest of the world as Tina Turner. Although a successful recording star, Turner found herself in an abusive marriage to a man who was important to her professional career. Neither chart-topping singles and albums nor the financial rewards that accompanied them could calm the storm that raged inside her. Instead of happiness and joy, her marriage to Ike Turner produced resentment, anger, and self-hatred. Then in 1974 a secretary for the Ike and Tina Turner Revue introduced Tina to Buddhism. She credits the meditation techniques she learned from reading and talking about Buddhism with giving her the inner strength necessary to take charge of her life. The peace she found in her new religion has been tapped by hundreds of thousands of black Americans in the last three decades.

Of the estimated 3 to 4 million Buddhists in America, roughly 70 to 75 percent are people of color. A significant number of those are African American women. Well-known black female Buddhists, like the author and cultural critic bell hooks, as well as Alice Walker, have shown black women that they do not have to sacrifice their blackness to reap the benefits of Buddhist wisdom traditions. At the same time, they can move beyond the externalities associated with race and religion in American culture by taking an inward approach to self-discovery. Angel Kyodo Williams notes, "Before, when [blacks] joined a movement—be it SNCC, NOW, or Greenpeace—we were trying to dismantle something outside ourselves, when we really needed first to have a revolution from within."

In the late 1960s, a new religion emerged that gave blacks the opportunity to find God right within themselves. The Nation of Gods and Earths, often referred to as the Five Percent Nation of Islam, or the Five Percenters, grew out of the Nation of Islam in New York City. Its founder, Father Allah (Clarence 13X), was once a member of the

Nation of Islam Temple 7 in Harlem. He was reprimanded on numerous occasions by mosque officials for questioning the divinity of Fard Muhammad. He persisted in challenging the black Muslims with a race-conscious question: If God came to earth to save black people, why would he choose to come in the form of a nonblack man? To him, deifying Fard was just one step away from worshiping a white man. Unable to get a satisfactory answer, he left the Nation of Islam and established a small, informal group that built upon some, but not all, of Elijah Muhammad's teachings.

Like the Nation of Islam, the Nation of Gods and Earths experienced rapid growth in American correctional facilities, which, by 1990, were teeming with young black men. While constituting just 11 percent of the nation's total population, by the end of the twentieth century black men and women made up nearly 50 percent of the state and federal prison population. Violent offenses by blacks had nearly doubled since the close of the Vietnam War, and it appeared that decades of benign neglect had resulted in a mood of nihilism in many black communities. Father Allah's movement provided ready answers to those behind bars and lost in urban decay.

Central to the ideology of the Nation of Gods and Earths is that it is not a religion at all but a natural way of life. To Five Percenters, a religion compartmentalizes obedience to the divine and does not offer a coherent plan for devotion in every area of one's life. Conversely, their science of Islam provides a code for living reflected in their spiritual philosophy of Supreme Mathematics. In this mystical scheme, each prime number and letter of the alphabet is assigned a value according to the teachings of Father Allah and his first disciples, the First Nine Born. In a manner similar to those of the Moorish Science Temple and the Nation of Islam, world geography is reconceptualized to reflect the Supreme Alphabet. In the United States,

Harlem is known as Mecca to members of the group, while South Bend, Indiana, is Supreme Born (S.B.), Pittsburgh is Power Born, and so on. Father Allah also taught his followers that every black man is a god whose proper name is Allah. The fact that God is a living man is reflected by the representation of the letters: A-arm, L-leg, L-leg, A-arm, H-head. Female members are known as Earths, mothers of the universe, and as such are to be protected and respected by the gods. Male and female members often change their names to reflect their awakening to a new identity. For example, a male member might go by the name Peace B. Allah, while women often take the title Queen or change their surnames to Earth.

Since each man is a god and master of his universe, there is little room for a strict code of moral conduct within the movement. How can anyone tell a god how he should live his life? Its lack of rules and democratic, decentralized structure are very appealing to young people who are "fishin' for religion." Moreover, its Afrocentric worldview builds self-esteem and self-respect in followers who sense that the society does not value them. Although often mistaken for gang members by law enforcement officials, Five Percenters espouse basic principles that are similar to those of other black organizations that preach self-determination. In the midst of some of its more esoteric beliefs, the Nation of Gods and Earths advocates three things. Their literature states:

1. National Consciousness: As a nation, black people are the first in existence and all other peoples derived from blacks in Africa. National consciousness is the awareness of the unique history and culture of black people and of the unequaled contributions they have made as the fathers and mothers of all civilization.

2. Community Control: The Five Percenters demand control of the educational, economic, political, media, and health institutions

of black communities. Community control flows naturally out of the Science of Life, which teaches that blacks are the Supreme Being and are solely in control of their own destiny.

3. Peace: Peace is the absence of confusion (chaos) and the absence of confusion is Order. Law and Order is the very foundation upon which the Science of Life rests. Supreme Mathematics is the Law and Order of the Universe, this is the Science of Islam, which is peace. The group's ultimate goal is to achieve peace for its members.

The ideals of national consciousness, community control, and peace resonate with the desires of the scores of black urban youth who are seeking greater control over their lives.

Five Percenters believe that 85 percent of the black population is lost and has no idea about its true identity, its divinity, and its proper place in the universe as the original people. Another 10 percent, the talented yet treacherous tenth, as the Five Percenters called them, know the truth but have sold out to the white establishment in exchange for the material security it offers. They work in concert with the system to prevent the 85 percent from coming to a correct understanding of the divine and racial order of things. Fortunately, though, Allah has called into being a nation of Poor Righteous Teachers, the Five Percent, in order to lead the lost into righteousness and purity. One of the key features of the Five Percenters is their loose and democratic structure. Members gather at informal meetings called universal parliaments, and any member can speak or introduce an idea or action he or she wants the group to reflect on. All gods and earths stand together as equal, and each individual is her or his own spiritual guide.

Some of the most prominent adherents to Five Percent theology may be found in the hip-hop communities in America. The explosion of hip-hop music, fashion, and dance in the 1980s provided an accessible medium through which gods and earths could spread their message of black nationalism, self-love, and Afrocentrism. Former female members like Queen Latifah and Erykah Badu gave the earths something to cheer about as they instructed black women on how to be strong in the face of adversity. And groups like the Poor Righteous Teachers, Brand Nubian, and Rakim were major voices in the movement as it was transmitted across America's airwaves. With recent releases like "The 18th Letter" and "Mystery (Who Is God?)," Rakim continues to be one of the prophetic voices for Five Percent hip-hop heads.

While the Five Percenters represent a relatively small movement within black religion, their philosophy is indicative of an attitudinal shift among black America's youth. Today they seek a religion that is rooted in their experiences, many of which are quite different from those of their parents. Events like the imprisonment of the Reverend Henry A. Lyons, former president of the National Baptist Convention, USA, Inc., for embezzling money meant to rebuild black churches destroyed in racially motivated acts of arson have left young blacks suspicious of organized religion. Instead of membership in a church, they crave a deeper spiritual awareness that makes them better human beings while providing roots in a world of eroding religiosity.

Pictures of Dr. King and the Kennedys that graced their grandparents' sitting rooms have been replaced by posters of Marcus Garvey and Malcolm X. The message is that self-determination, self-reliance, and self-love must precede any attempt to build bridges with other racial and ethnic groups. The crosses these young people were given by family members at confirmation and baptism have been replaced with ankhs and cowrie shells, symbolizing their desire to reclaim African faith traditions that predate Christianity. Yes, the Holy Bible still occupies a central role in their eclectic spirituality. But instead of commanding a spot in the center of the coffee table, it is on the bookshelf joined by copies of the Quran, the Upanishads, Elijah Muhammad's *How to Eat to Live,* and Kahlil Gibran's *The Prophet.* They seek wisdom and truth wherever it may reside, and refuse to limit themselves to that which worked for their forebears.

Despite this new mood among young black Americans at the close of the twentieth century, most of them were not prepared to turn their backs on the faith communities that had nurtured them. But they wanted more: a worship experience that reflected their energy and vitality rather than what they regarded as the often cold, lifeless congregations of their parents. How could they infuse the old-time spirituality of their ancestors with an animation that would attract new Christians to the fold? Slowly, the interests of today's youth have crept into black Christian worship. Fewer and fewer churches begin services with the traditional devotional period, during which deacons and church mothers would line hymns in the manner reminiscent of slave religion. The theological purpose of the devotional period was to set a tone of fear and trembling, so one might approach God humbly. Now "praise teams"—usually the most gifted singers from the church choir—establish a worship tone of jubilation and celebration.

This new movement in black Christian worship has been fueled by the influence of hip-hop music and culture on black sacred music. Just as the blues transformed church music in the 1930s and '40s, and rock and roll reshaped standard black hymns and anthems in the 1960s and '70s, hip-hop has blown wide the divide between older styles of gospel and contemporary expressions of the tradition. No one person has had a greater impact in this area than Kirk Franklin. Born in Fort Worth, Texas, Franklin was abandoned by his teenage mother and father when he was three. A distant aunt, Gertrude Franklin, took the boy in and provided the best life she could for him. Although sixty-four years old at the time, she worked tirelessly to steer him clear of the street life that preyed on so many young black men. She collected aluminum cans on the weekends to pay for his piano lessons, and at age four Kirk was already playing the instrument. Although he did not know it at the time, it was his aunt's faith in him as a child of God that sustained him through those years.

By the time he was seven years old, Franklin had been offered a recording contract by a Christian music label, but his aunt declined the offer, citing the child's age and his need to focus on school. She said she did not want him to be seduced by fame before he had developed the moral and spiritual character to understand that God deserved all the praise. Nonetheless, people recognized his talent and continued to offer him opportunities to share his gifts. At age eleven he was called to be minister of music at the Mt. Rose Baptist Church in Fort Worth, given the responsibility of directing the entire musical division of a congregation. But this involvement did not prevent Kirk from also answering the call of the streets. He smoked marijuana, drank alcohol, and flirted with the gang life that had become a fixture in his neighborhood. While in high school he fathered a baby, and it seemed that his life was on a trajectory not unlike those of most of the black men he knew. While Kirk was living with a foot in each world, his aunt was praying for him. She had done the best she could for him with what she had, and now it was time for her to turn him over to God. Only God could be the father he never had growing up. Only God could protect him from the dangers he would encounter while running with neighborhood toughs. Only God could make him see his life was worth living.

When he was fifteen a close friend was murdered. Reflecting on that tragedy, Franklin said, "That stopped me in my tracks." Languishing in despair, he decided to turn his attention to Jesus and the church. But even though he rededicated his life to Christ, Franklin was still intrigued by the sounds he'd come to love on the streets of

Fort Worth. He had a vision of bringing together the Christian message and hip-hop music. He wanted to provide a new image of the church for his generation. Franklin says that the misery quotient in American communities is so high that people are looking for something to fill the empty places in their spirits. They like the music of the hip-hop generation, and all he did was allow God to use him to deliver a new message. "The demand is so high, and gospel music supplies that demand. People are still lost, they are hurting," said Franklin. He added that as long as people are in pain there will be a need for the message of gospel. "Family members are dying of AIDS, family members are in prison, people want to hear something more than 'booties and Bentleys,' more than just a party type of attitude.... Gospel music addresses real issues. People need to hear something with substance, and that is gospel music."

In 1992 Franklin selected seventeen singers from the Fort Worth area and formed a gospel choir called the Family. Their 1993 CD release, *Kirk Franklin and the Family*, was a glowing success, selling over 2 million copies.

Some of Franklin's tracks borrow riffs from popular (and secular) black songs, including cuts like Parliament's "One Nation Under a Groove." This combination led *USA Today* to describe his 1998 *Nu Nation Project* "songs of heavenly praise swaddled in blankets of funk." His projects also mimic hip-hop recording structure, with interludes that connect the songs and give each CD a narrative feel. Franklin's latest singing group, 1NC (One Nation Crew), is a multicultural ensemble whose disparate racial and ethnic backgrounds are celebrated. "I've always considered gospel music to be a representation of our faith, not a definition of any one particular musical sound or style," Franklin explains. "The faith is the faith, and that's unchanging. Our thing is telling people about the love of Jesus Christ. Period."

The combination of potent lyrics, funky dance beats, and innovative music videos has brought Kirk Franklin's brand of hip-hop Christianity to mainstream America. His style is to incorporate Christianity and popular culture.

Franklin's music fits with another trend in black American churches—the megachurch movement. Although megachurches are not limited to blacks, some of the most recognizable personalities are black pastors. Known for their connections with charismatic traditions, as well as their embrace of wealth and their emphasis on "word ministries" (Bible-based instruction during worship), megachurch pastors use every medium to spread their brand of the good life. One of the figures at the forefront of this "prosperity theology" is Bishop Thomas D. Jakes, Sr.

Born on June 9, 1957, in South Charleston, West Virginia, T. D. Jakes graduated from Center Business College and held several jobs in corporate America before entering the ministry. In 1980 he became the part-time pastor of a ten-member storefront congregation in Montgomery, West Virginia, the Greater Emanuel Temple of Faith. Within two years the church body had grown enough to require his services full-time. One of the first tasks Jakes set for himself was to create programs that would reach those who needed to hear a word from on high. So he started *The Master's Plan*, a local radio show that informed listeners that God wanted them to enjoy the fruits of their labor not only in heaven but in this life also. In 1983 his first annual Back to the Bible Conference (now simply called the Bible Conference) put him at the forefront of the word ministries movement in black churches.

Jakes achieved national fame in 1993 with the publication of his popular sermon "Woman, Thou Art Loosed," geared toward women who had experienced various forms of abuse by men. That same year Trinity Broadcasting Network (TBN) and Black Entertainment Television (BET) began carrying a weekly television program featuring

Kirk Franklin's *The Nu Nation Project* was a major hit on both
R&B and contemporary Christian gospel music charts.

Kirk Franklin and his family arriving at the Essence Awards in 1999.

Jakes, *Get Ready with T. D. Jakes.* Three years later he moved his family, along with fifty other families, to Dallas, Texas. Today he is pastor of the 26,000-member Potter's House in Dallas, which certainly classifies Jakes as a megapastor. The sanctuary seats just over 8,000 and is equipped with computer jacks at every seat so parishioners can download sermons.

While he has enjoyed tremendous success, Jakes is not without critics. Some Christians believe his ministry is little more than prooftexting under the guise of teaching, pulling from the Bible passages that seem to support his gospel of wealth and prosperity. Others believe his emphasis on God's blessings manifested through conspicuous consumption is against the teachings of God. Regardless of whether one agrees with his teaching, one cannot argue with the fact that something in his message is speaking to the spiritual needs of millions of Christians, black and white, throughout the world. They tune in on radio and television each week and respond to seemingly constant pleas for funds. Jakes and other televangelists and megachurch pastors provide salve to a people who feel most mainstream churches no longer fill their needs. They are tired of being preached at in the traditional manner, with the sermon providing plenty of entertainment but little to sustain the believer through the week. They want to leave church feeling they have learned something about God's word and God's purpose for their lives.

The black religious experience defies monolithic interpretations. Black Americans can be found in all sorts of American religious organizations. Even though most blacks remain part of a Christian worshiping community, more are seeking alternatives to the church. Islam is the fastest-growing religion in the United States, and its numbers are rapidly expanding among African Americans as well. With even Louis Farrakhan, leader of the Nation of Islam, increasingly leaning toward a more orthodox practice, Islam promises to hold a significant place in the black religious imagination. Hinduism and Buddhism also continue to make inroads into black communities in the twenty-first century, as globalization narrows the gap between East and West. And as blacks seek to reclaim their African heritage, African-based religions like Candomblé and Santeria will still have a place on the black religious landscape.

Regardless of the particular faith, black religion will continue to be fueled by the shared and individual experiences of a people forged

The Congregation of the Oblate Sisters of Providence was founded in 1829 by
Elizabeth Clovis Lange, a refugee from Haiti. The order was
one of the first for African Americans in the United States.

in the caldron of cruelty and molded into a new people ready to meet
the challenges of a new millennium. It was not merely their own grit
and determination that saw them through the tragedies of Jim Crow
culture in the South and de facto segregation in the North; it was their
faith that their relatively insignificant sufferings and struggles were
somehow a part of God's larger plan for the world. Whether one believes
that black suffering is redemptive and works for the good of those who
love Jesus, or that the plight of blacks is a result of what Elijah Muhammad termed white devilish tricknology, it is clear that black Americans
have come "this far by faith," and that faith will remain when all else
has fallen away.

Black Christendom today is much more richly textured. While there are still more black Baptists and Methodists than any other denomination, blacks are continuing to trickle into predominantly white denominations. The crisis in urban public education has led some black families to send their children to parochial schools. They want the discipline and values-based education offered by Roman Catholic and Missouri Synod Lutheran schools, and as a result a new generation of black children have been exposed to and found a home in Christian traditions completely foreign to their parents.

CONCLUSION

What will the stories be of black religious life of the twenty-first century? Will they simply follow the traditions of the past, or will they attempt to discover new vistas of faith? Will blacks continue to be overwhelmingly Christian, or will we witness a larger shift toward religions of the East? Will Islam continue to have a significant role in the African American religious experience? As long as the United States is a place of unfulfilled dreams for substantial numbers of poor black people, separatist religious movements like the Nation of Islam will find a receptive audience. As the African American middle class continues to expand, they will also be exposed to a range of cultural options seemingly unheard of in their parents' generation. These new options will include Buddhism and Hinduism as well as other wisdom traditions rooted outside the church.

As for Christianity, today's black youth display less regard for denominational loyalty. Most young black people who grew up in a Baptist church haven't the slightest idea of what it means to be a Baptist. The theological points that distinguish their tradition from Episcopalians are unimportant to them. Their main reason for remaining a part of a particular worshiping community is not tradition but the fact that something there works for them. Maybe they like the preaching, or maybe they are drawn to the way the congregation is vocal on social issues in the community. Perhaps it is the music that gives them a more lively worship experience. Whatever it is, it has less to do with doctrinal difference or family history than with personal preference.

Moreover, the development of public policy will shape the way black Christianity serves it constituents. If faith-based initiatives continue to garner widespread public support, one can envision the social service work of black churches expanding as funding increases. The more churches are able to do for those in need, the more receptive those individuals will be to the church's message. Hence, the government might play an important role in stimulating growth in church communities that seem to have become irrelevant to the lives of many blacks.

Black Americans may choose one religious tradition over another—they may even invent new forms of worship—but they always remain a people of faith. At the start of a new century, black Americans still hold firm to that unyielding faith of their fathers and mothers. In the past, faith was the lifeline that gave African Americans the promise that God had the ability, when necessary, to part the sea and reveal the path to freedom. Today faith holds out comfort for black people still struggling for equality in the house they helped to build—the American nation. Faith is sometimes the only reason to carry on in the face of the extreme number of young black men in prison; broken schools that fail to prepare children to compete; irrational violence, often by black people against black people; and children being raised by unmarried, teenaged parents as families struggle to stay together despite the pressures of modern life. Only faith can hold back the tears when necessary. Only faith can inspire the dream of a better day. Faith and faith alone stands as a mighty sword to defend as well as a mighty arm of comfort and a mighty trumpet declaring to the world that this child, black, white, brown or any other color, is God's child. The story of black America is a story of faith fulfilled.

> *The apostles said to the Lord, "Make Our Faith Greater."*
> *The Lord answered, "If you had faith as big as a mustard seed,*
> *you could say to this mulberry tree, 'Pull yourself up by the*
> *roots and plant yourself in the sea!' And it would obey you."*
>
> —LUKE 17:5–6

SOURCES

1. "GOD HAS A HAND IN IT"

A number of books help establish the social context of the Vesey Rebellion. David Robertson's biography, *Denmark Vesey: The Buried History of America's Largest Slave Rebellion and the Man Who Led It* (Alfred A. Knopf, 2000), provides a delightful account of the world Vesey inhabited. Additionally, Peter Wood's *Black Majority: Negroes in Colonial South Carolina from 1670 through the Stono Rebellion* (W. W. Norton, 1996) gives a rich account of the interdependent relationship of enslaved blacks and white settlers in colonial South Carolina and how that relationship changed over time. We quoted quite extensively from *Designs Against Charleston: The Trial Record of the Denmark Vesey Slave Conspiracy of 1822* (University of North Carolina Press, 1999). The editor, Edward Pearson, offers a rich introductory analysis of the trial transcript. Also, we drew on Margaret Washington Creel's *A Peculiar People: Slave Religion and Community-Culture Among the Gullahs* (New York University Press, 1988) for background on African-based religions in the Carolinas.

Information on Richard Allen and the independent black church movement he helped ignite comes from his autobiography, *The Life Experience and Gospel Labors of the Rt. Rev. Richard Allen* (Abingdon, 1983). There has been a significant difference of opinion regarding his dating of some important events. Allen reports in his autobiography that the "Gallery Incident" occurred in 1787, whereas the historian Milton Sernett provides credible evidence to suggest that it could not have taken place before 1794. Most religious historians accept the 1794 date, while members of the AME Church understandably date this event in 1787. See Milton C. Sernett, *Black Religion and American Evangelicalism: White Protestants, Plantation Missions, and the Flowering of Negro Christianity, 1787–1865* (Scarecrow Press, 1975), as well as Albert J. Raboteau's *A Fire in the Bones* (Beacon, 1995).

2. THE PRINCE

Terry Alford's *Prince Among Slaves: The True Story of an African Prince Sold into Slavery in the American South* (Harcourt Brace Jovanovich, 1977) is the most complete narrative account of Ibrahima Abdul Rahman's life. The bulk of our information on Rahman is distilled from this book. We also relied on Sylviane Diouf's *Servants of Allah: African Muslims Enslaved in the Americas* (New York University Press, 1998) for details about Islam among enslaved Africans, as well as on the transcript of an interview with her. Chapter 1 of the book provides an excellent overview of slavery and Islamic law, and how Muslim slavery differed from European slave systems. For a good introduction to English ideas about slavery and their impact on the institution's development in the colonies, see Betty Wood, *The Origins of American Slavery: Freedom and Bondage in the English Colonies* (Hill & Wang, 1997). A good overview of the impact of David Walker's *Appeal* in both enslaved and free black communities can be found in *To Awaken My Afflicted Brethren: David Walker and the Antebellum Slave Resistance* (Pennsylvania State University Press, 1997) by Peter P. Hinks.

3. SPEAK TO MY HEART

We quoted extensively from Sojourner Truth's autobiographical account, *Narrative of Sojourner Truth, a Bondswoman of Olden Time: With a History of Her Labors and Correspondence Drawn from Her "Book of Life"* (Oxford University Press, 1991). In addition, there are a number of useful biographies of Sojourner Truth, a couple of which provided insightful background material. Most notably, we consulted Nell Irvin Painter's *Sojourner Truth: A Life, a Symbol* (W. W. Norton, 1996) and Carlton Maybee's *Sojourner Truth: Slave, Prophet, Legend* (New York University Press, 1995). We also quoted transcripts of Blackside interviews with Painter and Margaret Washington Creel.

Christopher Clark's *The Communitarian Moment: The Radical Challenge of the Northampton Association* (Cornell University Press, 1995) gave us a general overview of Truth's communitarian experience, while information on her time with Prophet Matthias came from *The Kingdom of Matthias: A Story of Sex and Salvation in Nineteenth-Century America* (Oxford University Press, 1994) by Paul Johnson. For information on antebellum black women preachers, we consulted *Sisters of the Spirit: Three Black Women's Autobiographies of the Nineteenth Century* (Indiana University Press, 1986), edited by William L. Andrews, as well as *Daughters of Thunder: Black Women Preachers and Their Sermons, 1850–1979* (Jossey-Bass, 1998), edited by Bettye Collier-Thomas.

4. "GOD IS A NEGRO"

Most of the background for this chapter comes from transcripts of interviews with the historians Sandy Dwayne Martin, Ed Redkey, and Stephen Angell, as well as transcripts of hour two of the television series *This Far by Faith*. Angell's *Bishop Henry McNeal Turner and African-American Religion in the South* (University of Tennessee Press, 1992) is indispensable for anyone interested in Turner and the AME Church in the South during Reconstruction. Nell Irvin Painter's *The Exodusters* (Alfred A. Knopf, 1977) helped us to understand the mass movement of blacks in the nineteenth century. The information on Lott Carey comes from Leroy Fitts's *A History of Black Baptists* (Broadman Press, 1985).

5. THE BUSINESS OF RELIGION

Most of the information in this chapter comes from Quinton Dixie's unpublished manuscript on Elias Camp Morris and the National Baptist Convention at the turn of the century. We also quoted from Morris's unpublished personal journal, in addition to the reprint version of his *Sermons, Addresses, Reminiscences, and Important Correspondence*. Bobby Lovett's *A Black Man's Dream: The First 100 Years: Richard Henry Boyd and the National Baptist Publishing Board* (Mega Corporation, 1993) was an important resource, as were Evelyn Higginbotham's *Righteous Discontent: The Women's Movement in the Black Baptist Church, 1880–1920* (Harvard University Press, 1993) and James Melvin Washington's crucial study *Frustrated Fellowship: The Black Baptist Quest for Social Power* (Mercer University Press, 1986).

6. "SAVED, BAPTIZED AND HOLY GHOST FILLED"

Quotations by Charles H. Mason are taken from *Bishop C. H. Mason and the Roots of the Church of God in Christ* by Ithiel Clemmons. We quoted E. C. Morris's address "Sanctification" in his book *Sermons, Addresses, Reminiscences, and Important Correspondence* to show the distinction between Baptist and Pentecostal views on holiness. Background information on Pentecostalism and the Azusa Street Revival comes from Vinson Synon, *The Holiness-Pentecostal Tradition: Charismatic Movements in the Twentieth Century* (W. B. Eerdmans Publishing Company, 1997). We derived details about the theological differences between Mason and Charles Price Jones from Dale Irvin's unpublished paper on Jones delivered at the Northeast Seminar for the Study of Black Religion and from *Saints in Exile: The Holiness-Pentecostal Experience in African American Religion and Culture* by Cheryl Sanders (Oxford University Press,

1996). The material on Father Divine comes from Jill Watts's *God, Harlem, U.S.A.: The Father Divine Story* (University of California Press, 1992).

7. BLACK GODS OF THE CITY

We quote W. D. Fard from Richard Turner's *Islam in the African American Experience*, and catechism material for the Moors comes directly from *The Holy Koran of the Moorish Science Temple*. We used biographical information on Fard and Elijah Muhammad as it appears in Claude Andrew Clegg III's *An Original Man: The Life and Times of Elijah Muhammad* (St. Martin's Press, 1997) and *The Messenger: The Rise and Fall of Elijah Muhammad* (Pantheon Books, 1999) by Carl Evanzz. *Message to the Black Man in America* (Secretarius M.E.M.P.S. Publications, 1997) offers an important glimpse into the religious and political views of Elijah Muhammad, while a clearer statement of the Nation of Islam's theological beliefs is found in Mattias Gardell's *In the Name of Elijah Muhammad: Louis Farrakhan and the Nation of Islam* (Duke University Press, 1996). For information on black Americans and Judaism, see *Black Zion: African American Religious Encounters with Judaism* (Oxford University Press, 2000), edited by Yvonne Chireau and Nathaniel Deutsch. Information on Marcus Garvey and the African Orthodox Church came from Randall K. Burkett's *Garveyism as a Religious Movement*.

A handful of books provided context for the transformation of black religion in urban areas. *Bound for the Promised Land: African American Religion and the Great Migration* (Duke University Press, 1997) by Milton Sernett best demonstrates the impact of rural-to-urban migration on mainstream black religious institutions. Michael W. Harris's *The Rise of the Gospel Blues: The Music of Thomas Andrew Dorsey in the Urban Church* (Oxford University Press, 1992) does a masterly job of explaining the migration's impact on black sacred music.

8. PRAYERS OF THE RIGHTEOUS

Transcripts of interviews with James Lawson provided the majority of the quotations in his section of this chapter. We also quoted Howard Thurman from his books *Jesus and the Disinherited* (Abingdon-Cokesbury Press, 1949) and *Footprints of a Dream* (Harper & Row, 1959). Background material on Thurman came from Walter Fluker and Catherine Tumber's edited collection of his writings titled *A Strange Freedom: The Best of Howard Thurman on Religious Experience and Public Life* (Beacon Press, 1998). A valuable resource for uncovering the relationship between Gandhi and black Americans is *Raising Up a Prophet: The African-American Encounter with Gandhi* (Beacon Press, 1992) by Sudarshan Kapur.

We quoted Martin Luther King, Jr., from *A Testament of Hope: The Essential Writings of Martin Luther King, Jr.* (HarperSanFrancisco, 1986), edited by James M. Washington. Other resources we consulted on King include Lewis Baldwin's texts, *There Is a Balm in Gilead: The Cultural Roots of Martin Luther King, Jr.* (Augsburg-Fortress, 1991) and *To Make the Wounded Whole: The Cultural Legacy of Martin Luther King, Jr.* (Augsburg-Fortress, 1992). We also consulted *I May Not Get There with You: The True Martin Luther King, Jr.* (The Free Press, 2000) by Michael Eric Dyson.

9. A CALL TO WITNESS

Much of the information on Fred Shuttlesworth and the bombing of Sixteenth Street Baptist Church, including Shuttlesworth quotations, was taken from *But for Birmingham: The Local and National Movements in the Civil Rights Struggle* (University of North Carolina Press, 1997). We relied on *Behind the Stained Glass: A History of Sixteenth Street Baptist Church* (Crane Hill Publishers, 1998) for historical context on the congregation and its role in Birmingham's black community. General information about the Civil Rights movement came from Taylor Branch's two books *Parting the Waters: America in the King Years, 1954–63* (Simon & Schuster, 1998) and *Pillar of Fire: America in the King Years, 1963–65* (Simon & Schuster, 1999).

William D. Booth's *A Call to Greatness: The Story of the Founding of the Progressive National Baptist Convention* (Brunswick Publishing Corporation, 2001) provides more than one hundred pages of documents relating to the crisis in the National Baptist Convention and its eventual schism. Quotations about the life and work of L. Venchael Booth came from this book. While there are a number of texts on King and the Civil Rights movement, the two that provided the best information on his organizational life are *To Redeem the Soul of America: The Southern Christian Leadership Conference and Martin Luther King, Jr.* (University of Georgia Press, 1987) and *The Origins of the Civil Rights Movement: Black Communities Organizing for Change* (The Free Press, 1984). Two chronologies gave us insights into racially motivated bombings and burnings. *My Soul Is a Witness: A Chronology of the Civil Rights Era, 1954–1965* (Henry Holt, 1999) deals with events related to the Civil Rights movement, while *Racial and Religious Violence in America: A Chronology* (Garland, 1991) deals with reports of violent incidents.

10. "THE BLACK MESSIAH"

Albert Cleage, Jr., has two books from which we quoted extensively. *The Black Messiah* (Africa World Press, 1989) is a collection of his sermons, while *Black*

Christian Nationalism: New Directions for the Black Church (Morrow, 1972) is an outline of his program for a movement of black churches committed to an Afrocentric theology. Also helpful was Hiley Ward's *Prophet of the Black Nation* (Pilgrim Press, 1969), which is the only full biography of Cleage to date.

Like Elijah Muhammad and the Nation of Islam, Malcolm X is receiving a fair amount of critical study. We relied on *The Autobiography of Malcolm X* (Grove Press, 1964) for perspective on his life before joining the Nation of Islam, while Lewis A. DeCaro, Jr.'s *On the Side of My People: A Religious Life of Malcolm X* (New York University Press, 1996) provides a rich portrait of Malcolm's spiritual journey. James H. Cone's important text *Martin and Malcolm and America: A Dream or a Nightmare* (Orbis Press, 1991) helped us locate Malcolm's views on civil rights and American politics in relation to those of King and Elijah Muhammad.

11. FISHIN' FOR RELIGION

The quotation from Charles Long comes from his essay "Passage and Prayer: The Origin of Religion in the Atlantic World," which appears in a volume edited by Quinton Hosford Dixie and Cornel West, *The Courage to Hope: From Black Suffering to Human Redemption* (Beacon Press, 1999). Additionally, a transcribed interview with West gave us some of the statistical data on black America in the 1970s, as well as background on the changed mood of young people during the period. For information on Buddhism in the black community, we quoted Angel Kyodo Williams from her book *Being Black: Zen and the Art of Living with Fearlessness and Grace* (Viking Compass, 2000), as well as newspaper articles on black women Buddhists. Mattias Gardell's *In the Name of Elijah Muhammad* was an important resource for understanding the development of the Nation of Gods and Earths, but the bulk of our material came from Five Percenter websites, especially www.allahsnation.net. This site is a part of the 5% Network, which is an assortment of websites representing the Nation of Gods and Earths across the United States and Canada.

Kirk Franklin and T. D. Jakes are constantly making headlines for their respective ministries, and we were able to get up-to-date information on them from newspapers, industry magazines, and websites. Franklin's autobiography, *Church Boy* (Word Publishing, 1998), is an account of his rise out of poverty and into stardom. Jakes's *Woman Thou Art Loosed* (Treasure House, 1993) helped us understand the power behind his conferences, which routinely draw 50,000 women.

ACKNOWLEDGMENTS

Dedicating this book to Henry Hampton only hints at the depth of feeling that his life generated in so many people. Henry touched souls— he helped, healed, encouraged, disciplined, and inspired. His life as a passionate visionary created landmark projects. His life as a businessman kept those projects afloat financially when others lacked the heart to hang tough. He gave a first break to so many filmmakers, film editors, and writers. He gave American history a tremendous boost by putting the story of African Americans into books and films to thrill generations. We are all his children.

Henry Ferris, the able and creative editor behind this book, joked that its title should really be "This Far by Luck." The book had to overcome hurdle after hurdle to become a reality, and the key to its success is the faith that Henry Ferris and Doe Coover, our agent, showed in the project at all times. Doe's steadfast dedication to her friend Henry Hampton and to the idea of this book was the light across stormy waters that eventually led us home. My friend Dante James gave his heart to this project and asked me to get involved. Thank you, Dante.

Special thanks to my family for putting up with a distracted, always busy dad. My wife, Delise, earns my eternal thanks for her daily support. Much love to Antonio, Regan, and Raphael.

I would also like to thank: Roger Ailes, Wally Ashby, Robert Barnett, Rachel Bebchick, Susan Bell, Douglas Bennett, Daniel Berger, Julian

Bond, James Brown, Darien Conklin, Gary Conklin, Ingrid Conklin, Kali Conklin, Richard Cooper, Joan Countryman, Bob Davis, Herbert Denton, Rick DiBella, Bruce Drake, Kay Edstene, Ken Eisenstein, Nancy Eisenstein, Ronald Elving, Susan Feeney, Shantelle Fowler, Vernon Francis, Deborah Frazer, Ofelia Garcia, David Garrow, Cheryl Gibert, Cheryl Hampton, Elaine Hansen, the Reverend John Harmon, William Harris, Love Henderson, Gigi Hinton, Molly Hooper, Brit Hume, Kim Hume, Alexandra Jenny, Beat Jenny, Elena Jenny, Jonathan Jenny, Stephen Klineberg, Kevin Klose, Catherine P. Koshland, Kathleen Larsen, Byron Lewis, Jr., Mara Liasson, Cynthiana Lightfoot, William Light-foot, James Loadholt, Jared Loadholt, Ted Love, Laurie Luhn, Howard Lutnick, Robert MacCrate, Ginger Macomber, Deborah Mathis, Ellen McDonnell, Gabe Mehretaab, John Moody, the Reverend Earl Neil, Robert Nevitt, Mary Lou O'Callaghan, Allegra Pawlowski, Veronica Petersen, Jackie Pham, Delsie Phillips, Jennifer Pond, Flip Porter, Gayle Potter, Pam Prue, Penny Prue, Barbara Rehm, Russell Reno, Carlos Rodriguez-Vidal, Marty Ryan, Benn C. Sah, Heather Salsbury, Harry Sandler, Deborah Lafer Scher, Dan Schlossber, Robert Schwartz, Jill Sherman, Eunita Simmons, Tony Snow, Perry Steinberg, Ken Stern, Francis Stokes, Arn Tellem, Robert Swift, Yolanda Tate, Jan Tavitian, Avery Teal, Brooke Teal, Christopher Teal, David L. Thomas, Jr., Diane Thomson, Lawrence Tint, Tom Tritton, Peter Trueblood, Roberto Waithe, Andrew Walker, Bob Walker, Jim Wallace, Delois Ward, Arleathia West, Arthur M. West, Arthur N. West, Chip West, Marisa West, Minna West, John Whitehead, Alma Williams, Antonio Williams, Armstrong Williams, Ashley Williams, Christopher Williams, Raphael Williams, Regan Williams, Roger Williams, Wes Williams, Audrey Wynn, and Barry Zubrow.

—JUAN WILLIAMS

I would like to acknowledge the hard work and patience of Doe Coover and the entire staff at the Doe Coover Literary Agency. Thank you for keeping the faith. I also thank our editor, Henry Ferris, for his stern and steady guidance. Without his expertise this book would read like a series of college lecture notes. Finally, my wife, Kimberly Dixie, put up with a great deal during the writing of this book and I want her to know that I appreciate her and all she has done for me.

—QUINTON DIXIE

In 1998, after the devastating loss of our brother, Henry Hampton, my sister, Veva Zimmerman, and I were given the challenging opportunity to manage Henry's beloved Blackside, a company he had started over thirty years before. In 1987 Blackside produced the now legendary documentary series *Eyes on the Prize*, a project Henry had nurtured for nearly twenty-five years before it made it to television. Henry was like that: a good idea was worth pursuing, even if one had to endure many setbacks and struggles and work for years to accomplish the goal. Time and trouble had met their match in our brother.

This Far by Faith didn't stray far from Henry's model. The idea of chronicling the impact of faith on community—so clearly integral to the civil rights story—was a natural outgrowth not only of *Eyes on the Prize* but of other work Blackside had done in the meantime. It would take years of fund-raising, planning, and, yes, sometimes stumbling before filming could actually begin. But Henry never once thought of turning back. It was an important story to tell, and a few setbacks were not going to deter him. And setbacks there were. But throughout the process, even after Henry's death, one thing was clear: This series was going to make it to television.

The television series and this companion book are the result of the inspiration and selfless work of many, many people, most of whom June

Cross has eloquently thanked in her notes. To these I would add June herself; as executive producer she helped rescue the television series when it seemed we might be one setback over our limit. I would also like to thank the authors of this book, Juan Williams and Quinton Dixie, who have so brilliantly translated the heart and soul of the series into print. They were ably assisted by the researcher Terrance Johnson. Doe Coover and her colleague Frances Kennedy moved mountains to help the publication of this book. Toby Greenberg sifted through hundreds of photographs to bring it to life and kept track of them all to boot. And Henry Ferris, our editor at William Morrow, never once lost the faith as he oversaw our journey from idea to finished book. On behalf of my sister and on behalf of Henry, I thank them all.

—JUDI HAMPTON

"The greater the source, the grander the river," goes an old Buddhist saying. A grand stream of talented and dedicated people—from researchers to producers, editors, filmmakers, and foundation officers—contributed their faith and deeds to realize this series and its companion book.

This Far by Faith perhaps marks the last of the epic Blackside documentary series, and the first finished bereft of Henry Hampton, Blackside's founder and its guiding light for the last thirty years. Henry, with help from Lulie Haddad and Martha Fowlkes, had largely shaped the series before he died. Martha had worked ceaselessly to secure funding. Terry Rockefeller had worked very hard on a proposal, favorably received, for the National Endowment for the Humanities. Then the chronic problems that plague all independent documentarians multiplied. Dante James stepped forward to shoulder the tremendous burden of executive producer of the television series, determined to move forward when paralysis seemed the only reasonable alternative.

Dante, Lulie, Callie Crossley and Jon Else laid the foundation for the series. Their work and the efforts of a research and development team that included Sheila Bernhard, Beverly Jackson, Meredith Woods, and Lisa Jones paved the way for the production.

A team of scholars, preachers, laypersons, and advisers gave unselfishly of their time and reminded us that the faith of African Americans finds abundant expression outside the walls of the church. Vincent Harding, our spiritual guide, took every opportunity to remind us that faith encompasses the total aesthetic, social, and political experience of a people, as well as the disparate religious rituals they practice. Dr. Horace Boyer gave unstintingly of his time and expertise to ensure that we got the period music right. Quinton Dixie, in addition to coauthoring this volume, provided invaluable advice to the producers as we sought to turn facts and stories into narrative. He and Juan Williams, an old friend of Blackside, have done a remarkable job with this book, aided at its inception by former Blackside staffers Jass Stewart and Marcus Walker. Dr. James Cone and W. W. Law not only shared their stories but made sure that we got the stories right.

Producers Noland Walker, June Cross, Lulie Haddad, Alice Marko-witz, Valerie Linson, and Leslie Farrell developed sensitive films that were beautifully realized by a team of talented editors. Jon Else and Callie Crossley gave precious time, energy, and deep thought to help us make cohesive, intelligent, watchable films. Our associate producers, working with a team of unpaid interns, managed to do the jobs of several people. Julia Elliot functioned as both a casting director and production coordinator; Lillian Baulding, Tanayi Seabrook, and Carol Bash multitasked their way through an office move on the eve of field production and deftly handled shifting deadlines; Sharon LaCruise organized both deliverables and the office body politic; and Kara Mathis took copious notes of our production meetings so that nothing would be forgotten. Bernard Jaffier and James Balls, pressed into service in ways they had least expected, contributed in equally unexpected ways. Turloch McDonough kept our computers and phones in working order. Cindy Kuhn and Charles Ellis found themselves the last ones standing when all others had surrendered hope. Their dedication left a trail for us to follow.

We are indebted to several people who went above and beyond the call of duty to ensure completion of this project. In this respect, Joy Thomas Moore at the Annie E. Casey Foundation deserves special recognition. Mabel Haddock at the National Black Programming Consortium, John Santos at the Ford Foundation, Alice Myatt at the Public Broadcasting Service, and Woody Wickham and Elspeth Revere at the John D. and Catherine T. MacArthur Foundation were especially supportive. Marita Rivero, our guardian angel at the WGBH Educational Foundation, refused to give up; and at the crucial hour Peter McGhee became our patron saint.

There were those who didn't complete the journey with us but on whose shoulders we nevertheless stand. Judy Richardson's long-expressed notion that Blackside's mission included an obligation to present the spirit of a people helped many of us keep our eyes on the prize, even after she left the company. Prathia Hall, braving a terminal illness, cooperated in the filming of one episode because she felt it was important to tell the faith stories of the foot soldiers in the Civil Rights movement. She did not live to see the series air.

In the beginning, of course, there was Henry, whose vision of telling the stories of everyday African Americans, in their own words, provided inspiration for so many producers and storytellers in Boston and across the country. This series began as an idea Henry had to tell the story of how black faith had sustained a people and changed this nation. "We'll do it for Henry," we told ourselves when things looked grim. We knew he would not have given up, and so neither did we. In the end we dedicate our efforts and work to his memory. Thank you, Henry, for all your inspiration over the years.

—JUNE CROSS AND THE PRODUCERS OF
This Far by Faith

ILLUSTRATION CREDITS

Page 10: Hulton Archive/Getty Images; 12: Culver Pictures; 16: LC; 20: SC; 21: SC; 39: Rare Book, Manuscript, and Special Collections Library, Duke University; 42: LC; 46 (top): SC; 46 (bottom): LC; 47: Amherst College Archives and Special Collections, by permission of the Trustees of Amherst College; 52–53: SC; 72: American Philosophical Society; 73: LC; 74: SC; 79: LC; 82: SC; 86: *Broadway, North from Canal Street*, drawn and etched by T. Horner, 1835, Museum of the City of New York, bequest by Mrs. J. Insley Blair, in memory of Mr. and Mrs. J. Insley Blair, 52.100.27; 88–89: Culver Pictures; 90: American Antiquarian Society; 92: LC; 95: SC; 97: LC; 100: SC; 105: The Metropolitan Museum of Art, Rogers Fund, 1942 (42.95.29); 108: SC; 116: LC; 119: Allen-Littlefield collection, Special Collections Division, Robert W. Woodruff Library, Emory University; 120: LC; 121: LC; 124: BC; 127: LC; 131: SC; 135: LC; 142: LC; 148: FPHC; 150: FPHC; 155: Courtesy Quinton Dixie; 156: FPHC; 157: FPHC; 162: FPHC; 169: AP/WW; 170: National Archives; 172: University of Illinois at Chicago, The University Library, Department of Special Collections, Arthur and Graham Aldis Papers, Aldis neg 1; 174: LC; 175: LC; 176: SC; 178: SC; 190: SC; 191: BC; 198: AP/WW; 209: Private Collection of the Reverend Jim Lawson, Jr.; 211: Hulton Archive/Getty Images; 213: AP/WW; 219: BC; 221: Ted Williams/CORBIS; 224: Donald Uhrbrock/TimePix; 225: Howard Sochurek/TimePix; 227: Howard Sochurek/TimePix; 230: Danny Lyon/Magnum Photos; 236: Bruce Davidson/Magnum Photos; 237 (top): BC; 237 (bottom): Leonard Freed/Magnum Photos; 238: BC; 243: BC; 247: BC; 254: Bob Black; 257: AP/WW; 261: AP/WW; 265: AP/WW; 269: BC; 279: Bowers Museum of Cultural Art/CORBIS; 282: Flip Schulke/CORBIS; 295: Steve Grayson/Getty Images; 296: Mitchell Gerber/CORBIS; 298: Dixie D. Vereen.

INDEX